Reaching the Solar Tipping Point

How Solar Thermal Farms, Photovoltaics and Electric Vehicles Will Transform Our Energy Future

First Published 2009
ISBN: 1-4392-3733-6

Library of Congress Control Number: 2009903633

Published by BookSurge
For more information visit
BookSurge at www.booksurge.com

03 10

Contents

Reaching the Solar Tipping Point

How Solar Thermal Farms, Photovoltaics and Electric Vehicles Will Transform Our Energy Future

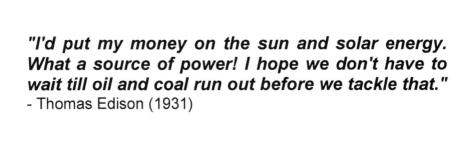

"I'd put my money on the sun and solar energy. What a source of power! I hope we don't have to wait till oil and coal run out before we tackle that."
- Thomas Edison (1931)

Introduction

What if there existed an energy production and transportation solution for the U.S. and other countries that was pollution and carbon-free and 100% forever renewable at a reasonable cost?

This book offers the reader a clear view of how the U.S. and other nations can achieve a significant portion of these goals through integrating solar thermal electric farms, photovoltaics, a new power grid and electric vehicle (EV) transportation. You will see how we can live largely free of fossil fuels for our direct energy and transportation needs. You will also see how cost-effective this solution will be in the future.

This book is based on 5 important attributes necessary for any fundamental large scale national energy solution.

Whatever your beliefs about the future risks of man-made CO_2 build-up in the atmosphere, it is clear that greater reliance on domestic clean renewable energy is good public policy.

An important aspect largely overlooked in the debate about atmospheric CO_2 build-up from fossil fuels is the possible affect of increasing CO_2 levels on our ocean's surface level pH (ocean acidification). It is possible that increasing CO_2 concentrations in the atmosphere may increase the upper 200 feet of our ocean's pH balance sufficiently to adversely affect calcium carbonate shell formation in corals, mollusks and hard-shelled invertebrates. The risk to the ocean's pH balance is currently not well understood.

This book will also show you how you can make choices to be part of an intelligent, safe, permanent clean renewable energy solution—and how to encourage policy makers and regulators to adopt elements of this roadmap. You will learn about the amazing revolution taking place with solar technologies and electric vehicles.

You will also see why some existing policy choices like nuclear power and food crop biofuels have significant adverse long term impacts. You will clearly see why the hype behind corn ethanol and cellulosic ethanol is really a false mirage and will result in wasted taxpayer dollars and even more pollution.

A modern, high efficiency solar thermal farm for electricity generation is a proven technology that can provide day and night electricity generation. Combining solar thermal farms with photovoltaics and electric and plug-in hybrid vehicles (PHEV) can provide us with the freedom to live and drive affordably and nearly carbon fuel free. We have a sunlight fuel source that is endless and non-polluting. Consume as much solar generated electricity as you like guilt-free, while saving more then 80% on transportation fuel costs compared with gasoline or diesel. America and other nations will once again embrace unlimited driving without regard to consuming imported fossil fuels or polluting the air while spending 70-80% less on renewable fuel.

Wind power will also play an increasing role in our future, but wind power is the small sibling to solar power. Wind power is limited in its reliable scale in the United States and lacks consistent availability for large scale base load electric generation capacity. Wind power will however be highly complementary to a solar electricity power infrastructure, and along with geothermal power, will add diversity to the renewable energy mix.

The sheer scale and consistent availability of solar electricity dwarfs what is available from wind power and other renewables in the United States. Solar thermal farms in the U.S. combined with overnight thermal storage will cover only a small fraction of our uninhabited desert and other arid land areas. These solar farms will be capable of providing enough power for much of our U.S. electricity and ground transportation needs. Add rooftop photovoltaics and photovoltaic farms and you have a compelling large scale efficient source of 24 hour per day electric power.

The cost of solar thermal farms is not much more then the current cost of natural gas generated electricity, and may soon approach the cost of increasingly expensive and regulated coal generated electricity without the environmentally damaging side effects. Once the solar installations are completed, solar electricity produces zero pollution, CO_2 or other greenhouse gases.

The technology for a new, long distance high voltage DC (HVDC) electric grid backbone needed to carry the power from the solar farms to most parts of the country will be described in this book. The technology for HVDC has been proven in high power commercial applications for over 50 years in over 100 locations around the world.

You will also learn why so many major drawbacks from biofuel policies are hidden from the public or rarely discussed in the press. Cellulosic ethanol, for example, held out as the holy grail of the biofuels revolution, is at least seven to ten years from any practical commercial large scale development. It will also be a cause of air pollution, and will require 100 acres of cellulosic biofuel crop farmland to equal the electric transportation fuel equivalent generated from only one acre of a solar thermal farm. This equivalent one acre of solar thermal farm can also provide power for electric cars 24 hours a day. Six acres of solar thermal farm can provide the same transportation energy as one square mile (640 acres) of cellulosic biofuel crop land at a fraction of the cost over the decades-long life of a solar farm.

Unpublicized nonpartisan studies show that for cellulosic ethanol to begin to replace oil in the Unites States, nearly 100% of America's farmland would need to be consumed growing biofuel crops by 2050, whereas for solar farms, less then 1% of the equivalent land area would be required. The land used by the solar thermal farms would be primarily uninhabited deserts and arid areas not used for crop production. Yet ethanol currently receives the vast majority of renewable subsidies.

Our energy crisis has been a long time in the making. Societies globally are facing energy policy decisions that will affect their lives and energy security for the next several decades. Sometimes crisis leads only to adversity. But often crises can force positive changes too. As we move deeper into the 21st century, most countries are facing important decisions about their energy needs and food security—and for many countries, the dire need to raise their citizens out of poverty.

Virtually every day there are headlines in newspapers, TV and on the internet about our energy future, our economy, food supply, and questions about greenhouse gas uncertainty. We no longer have the luxury or time for misguided energy policies. We need to start making the right intelligent choices now.

The forces driving our energy and economic uncertainty and rapidly growing CO_2 emissions are really all part of the same globalization phenomenon. They have also been aggravated by very ill-informed policy decisions and special interest influence.

Many people, including well respected scientists, policy makers and journalists have even begun to proclaim the end of society as we know it and the end of our high standard of living as our energy and economic insecurity increases.

Fortunately you will learn in the following chapters the details of a safer, more reliable and cleaner energy future that is on the horizon. This will not happen overnight. We will also be dependent on continued investments in domestic and international oil and gas production for the next several decades in order to sustain and fund the transition to a more secure renewable energy future.

There will be dislocations and impacts on society's standard of living as the long term affects of peak oil and energy uncertainty ripples through the global economy. It is also true that the seeds of the most recent energy crisis are planting the opportunity for all of us to enter a new phase of decreasing foreign energy dependence, cleaner air and water, and eventually lower food costs for almost everyone. And part of the solution is as always rooted in the ingenuity of our planet's people and their creativity. It is also rooted in what the ancient Egyptians always knew to be the ultimate source and giver of life--Ra, the sun god.

But there is a very large risk to this better, safer and cleaner energy environmental future for everyone. This risk is rooted in the age old roadblocks of entrenched special interest groups and people not fully understanding the best choices that are available to them. You will see where the risks are greatest and how your choices can help society move to a cleaner, safer, and sustainable world for generations to come.

One of the surprises you may find in this book is that the individual elements and technologies of an integrated power and transportation solar electric solution are already at hand. They are plain to see once they are all linked together to paint a clearer picture of how they are interconnected.

I hope this book can play a part in informing you how solar energy and electric vehicles can dominate our energy future in the 21st Century. We can have clean affordable energy and transportation fuel without practical limit and eventually little dependence on foreign oil into the far distant future.

1

Our Energy Crisis- Seeds for a Brighter Future

It is no coincidence that at the same time China and India are endeavoring to bring their populations into the middle class and develop a consumer oriented society, we are seeing the limits of peak oil. Nor is it entirely coincidental that concerns about our energy future and CO_2 build-up in the atmosphere are both in the news at the same time. Matt Simmons' book in 2005, *Twilight in the Desert*, foretold what will be in store for the global limits of oil production and ever increasing oil reservoir depletion.

Since *Twilight in the Desert* was published, Peak Oil has received much media coverage. Two of the central points in Mr. Simmons' book are: (1) because of fewer large oil field discoveries in the last 30 years, the accelerating depletion rate from existing oil fields will make it far more difficult for overall global oil production to grow and may soon enter an irreversible decline, and (2) the Saudi oil fields are also suffering accelerating depletion, limiting Saudi Arabia's ability to grow oil production to offset production declines in the rest of the world.

Only a few years ago both the Saudi government and agencies like the International Energy Agency (IEA) projected that

Saudi Arabia could lift oil production in future years to 20 million barrels per day. Saudi production has fluctuated over the past several years between 8.5 and 10 million barrels per day. The Saudi government dismissed Matt Simmons' book as naïve. The intervening years have shown Mr. Simmons correctly identified a grim reality- the Saudi government is facing accelerating depletion and they have been unwilling to share the oil production data that would have alerted the world far sooner to the true limits of Saudi Arabia's oil production capacity.

In 2005, Saudi Arabia announced planned capital expenditures of $12 billion to increase oil production to 12.5 million barrels per day. In 2007 Saudi Arabia quietly announced planned capital expenditures of $65 billion to raise oil production to 12.5 million barrels per day, yet the international news media failed to note the enormous upward revision in capital expenditures to achieve the same projected oil production level announced 2 years earlier. Clearly the cost of maintaining oil production was skyrocketing for the world's largest holder of conventional oil reserves.

When increasing rates of oil field depletion, even in the oil rich Middle East, meet increasing global demand, ultimately the laws of supply and demand prevail. Higher oil and gasoline prices were the result. Economic recessions may provide temporary oil supply relief, but only for the short term. Rapid movement to other transportation fuels and higher fuel efficiency will be required to offset accelerating conventional oil resource depletion.

When another 2 billion people decide it's time to start on the path to driving cars and turning on the lights for their new apartments and houses, are we really surprised by the daily headlines heralding the dangers of Peak Oil? The impact of these global forces will be with us over the next few decades and the transition of our economy away from foreign oil dependence will be painful for many people and industries during this transition.

The current oil and coal based global economy was destined to hit a production supply wall. It was only a matter of when and how hard. In all likelihood that question has been answered. Most people don't really care how their oil, gas and electricity get to them, as long as it's reasonably cheap and you don't have to wait in line. The days of *long term* cheap oil and gas are over. Social and economic costs are straining the budgets of consumers, industry and governments. Greenhouse gas concerns add to the unease about increasing emissions from fossil fuels.

The Beginnings of Our Carbon Fuel Dependence

Mankind relied on wood for millennia as a primary energy source. It worked well for the needs of hunter-gatherers to stay warm, provide light, and cook food. Wood is a carbon based energy source that stores energy for up to a few hundred years, and it is easily harvested and burned.

In seemingly endless forests around the world, tribal peoples found easy access to abundant and available fuel supplies. Industrialization outstripped the easy access to readily available wood sources in the ever increasing quantities needed for the industrial revolution.

As the energy needs of societies entering the industrial revolution grew, coal was found to be available in the quantities needed to continue to feed the ever growing appetite of industrializing economies. Coal heated homes, powered industry and became a convenient transportation fuel as trains began to supplant horses and stock animals for longer distance travel. Coal was also easily mined and in proximity to where it was needed in many parts of the world. As train and ocean transportation evolved and improved, it became easier to move wood and coal even greater distances to fuel factory floors and homes far away from where coal and wood was harvested or mined.

As America industrialized, ever larger amounts of energy were needed as well as more convenient forms of fuel for transportation, heating and lighting. It was at this time in the later 1800's in America that oil quickly grew in importance.

Oil flows from natural springs had been used by Native American Indians for thousands of years. In western Pennsylvania oil was used by the Seneca Indians who used it for medicinal purposes. Oil wells were first drilled in western Pennsylvania by Edwin Drake in 1859. The first well was 69 feet deep and produced 15 barrels a day. Oil use after 1859 grew rapidly.

John D. Rockefeller's Standard Oil Trust was one of the most powerful industrial organizations to capitalize on the world's increasing use and need for more oil. Rockefeller's Trust controlled much of the production, transport, refining, and marketing of petroleum products in the United States and many other countries.

Originally, Standard Oil focused on making money in the home lamp and lighting market which was converting from whale oil and coal oil to kerosene. The emergence of the automobile and the need for gasoline, which at the time was a nearly worthless refined by-product, brought vast wealth to Rockefeller and his shareholders.

The world had discovered a new convenient transportable fuel and the race was on in far flung parts of the globe to tap this new energy source. Whole industries were built around the production, refining and use of oil and gas. Oil also became much more then a convenient liquid energy source. It soon became the raw material for a whole range of new materials, including plastics, asphalts, medicines, clothes, waxes, and chemicals.

The United States and the world have seen many episodes of oil booms, busts and panics in the more than one hundred years of oil production.

World War I saw the beginnings of concern regarding U.S. oil supplies. In the fall of 1918, states east of the Mississippi even saw a total of seven "gasless Sundays" put into effect because of the heavy demand for oil oversees.

Oil consumption in the U.S. grew rapidly after WWI, doubling the demand for oil from 1914 to 1920. From 1920 to 1922 the U.S. became a large importer of foreign oil and began a national policy of encouraging domestic oil companies to develop oil resources overseas out of fears of limited domestic oil reserves.

In 1919, the magazine *Scientific American* noted that the auto industry could no longer ignore the fact that only 20 years worth of U.S. oil was left. "The burden falls upon the engine. It must adapt itself to less volatile fuel, and it must be made to burn the fuel with less waste.... Automotive engineers must turn their thoughts away from questions of speed and weight... and comfort and endurance, to avert what will turn out to be a calamity, seriously disorganizing an indispensable system of transportation." As more oil resources were discovered in the U.S. and abroad, the concerns about oil shortages and conservation abated.

The production of oil worldwide rose substantially for the next 80 years with increasing discoveries matched by ever increasing consumption, primarily for use as a transportation fuel.

OIL AND GAS LIQUIDS
2004 Scenario

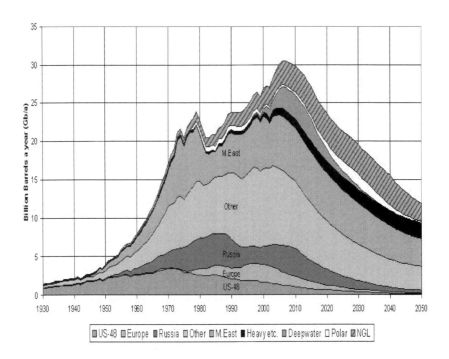

Uppsala Hydrocarbon Depletion Study Group
OIL AND GAS LIQUIDS 2004 Scenario

In 1956, the geologist M. King Hubbert, while working for Shell Oil, developed a model to describe and predict the rate at which oil fields typically produce oil over time. He predicted with reasonable accuracy that the United States would reach its peak oil production in 1971 and that a subsequent unavoidable decline would follow. Hubbert's "Peak Oil" theory has become the centerpiece of a long and continuing debate about when the world will slip into an irreversible decline in total world oil production. His model predicts that when an oil field has produced approximately 50% of the reserves contained in that field, the field falls into an irreversible production decline following a predictable rate of decreasing production.

Complicating the issue of peak oil are other unconventional sources of oil such as the tar sands in Alberta, Canada, and how to account for natural gas and natural gas liquids. It is becoming clear that whether or not the world has reached its absolute peak oil production capacity, the ever growing demand for oil and energy from Asia and other developing economies makes the question moot. It is clear the demand for fossil fuels has overwhelmed the oil industry's ability to increase the supply to these newly industrializing economies. Global recessions will only delay the inevitability of limited supply.

Ultimately all that is left to balance supply and demand are higher energy prices which act to destroy new demand, and the economic cycles that can cause a temporary lessening of demand until the next economic upturn. Because of increasing globalization, there has been a steadily increasing demand for fossil fuels independent of United States economic conditions. Net demand for oil has been decreasing in the United States while total global demand has continued to increase long term. The United States is no longer the driving economy for oil demand and this was reflected in the high peak prices seen in oil in 2007-2008 due primarily to increased Asian energy demand. As new transportation fuels and efficiency increases, we will begin to see a downward pressure on oil and gasoline prices, allowing the world economies to grow while global oil production reaches its peak oil limit.

Hydro-power has also for centuries been another, more limited source of cheap and convenient power for machines and eventually electricity. Hydroelectric power from rivers and dams is an excellent source of non-polluting electricity. Unfortunately, there isn't nearly enough supply of hydroelectric power to meet the large and growing needs of our industrial society.

With the advent of WWII and the development of nuclear technology, another new source of technologically available and potentially cheap energy came into its own. Originally sold to the world as a safe and inexhaustible source of electricity, the truth about nuclear power inevitably leaked and then exploded into the public reality of what the consequences are when things go wrong with nuclear power. The nuclear power industry originally promised the United States that a nuclear power plant accident was almost impossible or at least not likely for the next 10,000 years. The Three Mile Island nuclear reactor changed that timeline forever.

The Three Mile Island nuclear power plant was originally commissioned in September, 1974 and proclaimed to the public as a safe, reliable nuclear generating facility. The 10,000 year accident free estimate was a bit off the mark. A mere five years later, on March 28, 1979, the Three Mile Island nuclear facility came close to becoming by far the largest disaster in American history and potentially contaminating a large part of south eastern Pennsylvania for centuries, if not longer. The accident at Three Mile Island may not have been inevitable. But the nuclear waste it and all other nuclear power plants create is inevitable. Chernobyl was an even bigger real as well as public relations disaster for the nuclear power industry. The chapter on the nuclear power option addresses the question, "why would a society risk picking nuclear power as an energy source given that an accident or terrorist strike can have adverse consequences to people living in the area for thousands of years?"

Oil in the 1970's and 1980's witnessed several steep price rises followed by major price declines. The price shocks of the 1970's, 1980's and 1990-1991 were mainly due to oil supply disruptions. Beginning in 1980, the world began a continuing trend of consuming more oil then was being discovered. Because it takes years to move discoveries into production and years to increase production in large oil fields, the effects of this shortfall were not felt for more then two decades.

As the trend of fewer and fewer large oil field discoveries continued into the present, the effects of fewer oil discoveries, increasing oil field depletion and rising demand from emerging markets has combined to drive oil prices up over ten times the price of oil in the late 1990's. The average price of crude oil in 1998 was $11.91 per barrel. The price of crude oil in 2008 surpassed $145 per barrel for the first time.

Coal is seeing some of the same supply-demand limitations largely driven by the increasing industrialization of China and India. China had been a net exporter of coal, but in mid-2007 began to imported coal for the first time. Coal prices rose in 2008 to above $125 per metric ton. In 2003, coal in China traded at about $25 per metric ton.

As the limits of higher consumption rates of existing oil and other fossil fuel sources are being reached and combined with concerns about increasing greenhouse gas emissions, we see the results of public energy policy run amok. The full range of possible new sources of energy for transportation fuel, electricity and other uses has recently exploded into the public conscience and public policy debate.

Investments began flowing into many alternative and renewable energy programs with the added impetus of subsidies and incentives. Chief among the offspring of the new energy revolution are biofuels, primarily corn ethanol in the United States and sugar cane ethanol in Brazil. What has happened with the rush to biofuels reveals much about what is broken with public policy decision making, at least in the United States (and by all appearances, many other countries as well).

America's biofuels policy is a perfect example of how special interests have combined with the media's "green is good" philosophy which has led to a very predictable waste of taxpayer money and higher costs for global food and grain prices. Lost in the public policy debate is something rather important to consider- the old medical adage, "...at least, do no harm", widely attributed to Hippocrates.

Wind, solar and other forms of renewable energy are now seeing very large investment flows and dynamic innovations in technology, engineering and creative financial arrangements.

The limits of fossil fuel supplies and the adverse impacts of foreign oil dependency are painfully apparent to virtually everyone. But what to do about it? Even though there is still debate about the ultimate impact of rapidly increasing levels of CO_2 in the atmosphere, it is clear that burning ever larger quantities of fossil fuels is raising the concentration of CO_2 in the earth's atmosphere; with the risk of unknown long term side effects.

As mentioned previously, one side effect little discussed in the media is the impact of higher CO_2 levels on the ocean's pH. The increase in ocean surface water acidity (ocean acidification) is widely understood to potentially interfere with a sea crustacean's ability to grow a hard calcified shell.   Would oyster and clam lovers really wish to be deprived of healthy shellfish? The world's hard corals which make up the great oceanic reefs are also potentially in danger.

Much has been written regarding the science and evidence for man-made global warming and the potential risks to the climate this century as more fossil fuels are burned.   There are also credible scientists who dispute the scale and nature of the risks of increasing CO_2 levels on climate change. It is clear, however, that there is significant uncertainty from a risk mitigation perspective and that it is prudent to reduce or reverse the growth of CO_2 emissions in the long term.

The combination of peak oil and fossil fuel limitations on global economic growth have collided with the desire for risk mitigation of greenhouse gas emissions to create a global shift in the need for reliable, clean, affordable energy alternatives.

There are in reality few fundamental sources of energy for use on earth.   There are the long term stored sources, which include geothermal, fossil fuels, and energy stored in the orbital interaction between the moon, earth and sun which is expressed in the form of tidal energy.   There is also the only truly renewing source of energy, sunlight, which drives wind power, hydro-electric through water evaporation and rain, and all the biofuels which even includes algae.   Lastly there are the nuclear sources which include fission and fusion power.

The nuclear power industry has mastered the production of power using uranium in the core of power plants to produce heat which is then converted into electricity. There is also fusion power which is the process that drives the sun and stars and gives light to the universe. Several countries including the U.S. have mastered fusion power for the production of nuclear weapons, referred to as H- bombs, or Hydrogen bombs. Nuclear fusion is the process of fusing hydrogen into helium which liberates tremendous energy, but requires enormous pressure and temperatures.

In all likelihood, nuclear fusion will be our ultimate power source in the future. It is however, not a technology that will mature for power generation in the next several decades.

There is a shift underway from the stored fossil fuels and existing hydro-electric power sources to renewables, nuclear fission power, and to a very limited degree, tidal and ocean power. It is important to understand that all forms of renewables derive their source of energy from the sun and the sunlight that shines on this planet.

This book addresses why direct solar energy conversion is the most efficient, safest, cleanest and likely the most economical power source in the long run to harness the heat and electricity needed to power our future in the U.S. and those countries blessed with abundant and consistent sunlight.

Electricity is a superbly efficient form of energy to transmit long distances and turn into useful work such as lighting, motor driven applications and transportation through electric vehicles (EV's). The use of solar generated electricity in direct electricity driven applications such as lighting and air conditioning requires very few regulatory changes beyond the issues of long range transmission, land use and thermal storage which will be discussed in the following chapters. You will see that solutions to the long range transmission and storage needed for solar generated electricity already exist and are already in operation in many locations.

For EV's, solar generated electricity can be used directly to charge an EV battery used in a pure electric vehicle or as the battery in a plug-in hybrid vehicle. Electricity can also be converted into hydrogen, other liquid or gaseous fuels, or even drive compressed air cars. Many of these different EV options will also be described in the following chapters. A very important aspect of the benefits of electric vehicles, or vehicles that use hydrogen or compressed air is that there is no CO_2 generated from the vehicles and no air, ground or water pollution. Once it becomes clear to the public that even large cities can have their ground transportation needs met with reliable electric fuel, public support for increasing renewable electricity led by solar, and electric transportation, will become even more widespread.

The United States faces no real practical limits to growing its long term energy usage through the growth of solar thermal farms and photovoltaics, augmented with wind power. The transition costs will however be substantial. You will discover through the course of this book a clear path to a solar energy future for our electrical and ground transportation needs. Compromises will need to be made in the short and midterm as the electric vehicle technologies and choices are developed and infrastructure expanded over the next several decades. But you will see the elegant practicality of living from pure solar energy and driving cars, buses and trucks that are powered by sunlight generated electricity.

2

Energy Solutions- The Five Attributes

Remember the old humorous adage, "Ready, Shoot, Aim"? That's where we appear to be headed in the great debate and actions regarding energy policy around the world, and unfortunately in America too. We are in danger of losing our way.

One of the first and foremost questions that should be driving energy policy and the search for energy solutions is, "What are the required attributes of a long term sustainable energy policy?" The question applies to the U.S., as well as for each country faced with tough energy policy choices. This chapter addresses the attributes of an intelligent energy policy for the U.S. and how they drive the potential mix of energy sources and solutions. It will become evident as you go through this chapter that there has been a complete failure to address this fundamental question.

It is important to understand that oil, natural gas, coal, and existing nuclear power will be needed for the next 30-50 years during the transition to other primary energy sources. It is also important to understand that the U.S. needs to limit as much as possible its dependence on foreign oil through domestic production during this transition time. The U.S. will not be able to sustain the enormous trade deficit that has arisen due to its dependence on foreign imported oil without major adverse impacts to the economy.

Base Load Power

Base load power is the power source that provides foundation power 24 hours a day, seven days a week for a given power application. Currently in most regions of the United States, the base load power source for electricity is coal. In some regions it is natural gas, and in parts of the Pacific Northwest and countries like Sweden it is largely hydro-electric power. Base load power in the context of transportation is derived from oil in virtually all countries. At the present time, the use of oil as the primary fuel source for transportation exceeds 97% worldwide.

There are five fundamental attributes needed to assess the suitability of base load energy sources for a given power application. It should be understood that there will always be a mix of supportive energy sources for any country's base load power. You will see through the remainder of the book, how solar electricity can become one of the largest components of base load electric power and a primary transportation fuel, and in the longer term, the largest component for both. You will also see how solar electricity is capable of meeting our energy needs for the next century and beyond.

First, all future base load energy solutions need to be (1) reliable and (2) safe (including environmentally) both in the short and long term. The potential solutions need to have (3) sufficient scale to actually satisfy the national power needs and have room for growth. The solutions need to be (4) economical in the medium and long term. Lastly, the solutions should (5) not have unintended consequences or compete for resources in short supply or critical to the health and security of people dependent upon these resources.   Any unintended consequences need to be understood and accounted for in the overall policy choices.

Once we have sets of solutions that satisfy these 5 criteria: Reliable, Safe, Scalable, Economic, and No unmanageable adverse consequences, then we can go on to compare potential solutions to see which ones make the most sense for a given region, country or population. We can expect, of course, that some regions will have different solutions or mixes of solutions based on the particular resources and needs of the region.

Now let's take a look at the main candidates that are the current policy choices for the U.S. (and many other countries). First let us look at biofuels and how they stack up. We'll start with corn-based biofuels and then move on to other biofuel feedstocks.

Are corn-based biofuels reliable and safe? Surprisingly the answer is no! How can this be?

Haven't we been told a large part of our future energy mix for transportation should be based on corn ethanol? Billions of dollars in subsidies are fueling a large build-out of corn ethanol refineries. Entire infrastructures are being built for corn ethanol.

One of the primary reasons corn-based ethanol is not reliable is that food crops are always at risk from droughts, floods and diseases. It's been true throughout history and will always be true no matter how cleverly a bio-engineering seed company creates a new seed strain. Why would a country pick a critical transportation fuel source that may have a crop failure for one or more years? We'll address that question and other downsides to corn ethanol in the biofuels chapter.

Even worse, corn is a staple for our food supply. As a fuel source that serves two markets at the same time, if there is a crop failure, we not only face food shortages, but also fuel shortages for at least the next year's crop.

What about nuclear power? This is the saddest of all the energy option choices that has already been made in this country and other countries around the world. This choice has many of the worse energy attributes. First, is it safe? Secondly, what do you do with the nuclear waste? After more then 40 years of nuclear power generation there is still no safe way to eliminate nuclear waste. Nuclear power plants don't last forever. What happens to them when we de-commission them? The power plant core and nuclear fuel waste will be radioactive for many thousands of years. Who pays for the decommissioning and waste oversight in the far future? In theory the rate payers and nuclear power utility covers the cost. In reality your descendants will pay in dollars and health risks. Besides the risk of nuclear terrorism and risk of accidents, nuclear leaks and disposal costs, what happens if there is a major accident? It's also a short term solution that generates dangerous nuclear waste for well over 100,000 years. Other risks for nuclear will be addressed in the chapter on nuclear power.

What about coal? Existing technologies that burn or convert coal for fuel generate tremendous amounts of CO_2. The price of coal has also risen indicating we are beginning to reach the point where even cheap coal is becoming a thing of the past. It's becoming increasing unlikely that the world will tolerant ever more coal power plants or coal-to-liquid fuel plants producing ever larger amounts of CO_2. CO_2 sequestration, whereby CO_2 is piped to an underground depository, is currently uneconomical and not practical for a national economy the size of the United States. Clean coal is no better then putting lipstick on a pig- its still pork. Coal appears to have more then one Achilles' heel regarding future use; it isn't particularly clean to mine, ship, process, or burn.

Could we perhaps just go out and find more oil? The short answer is no. The world's inexpensive oil is being rapidly depleted and the needs of growing Asian populations are rapidly outstripping the ability to grow the supply of future oil production. The world's older oilfields are being depleted rapidly and new oil discoveries and production will not keep pace with new demands for more oil. Unconventional sources of oil, natural gas, and natural gas liquids simply can't supply what's needed in sufficiently increasing quantities without other new transportation energy sources.

Fortunately, higher oil and natural gas prices ultimately create their own replacements through demand destruction and alternatives becoming more competitive. The economies of countries that have surplus oil to sell will gain wealth and countries that continue to depend on large imports of oil, like the U.S., will see their economies suffer until they find intelligent alternatives to oil.

There are many pundits, energy policy decision makers and journalists who claim we will need to permanently change the way we live to accommodate the limits of our energy usage. If this assertion is true, so be it. But is it true? Fortunately, as you will see, the answer is no. The truth is that in the future there will be plenty of cheap, reliable, clean and safe energy that does not compete for other essential resources. It has all the attributes of what's needed for a long term affordable set of energy solutions, at least for the United States, and many other countries.

The solution can cover much of our energy needs for electricity, as well as transportation, and if needed, heating as well. It's non-polluting and will never be exhausted. In short, once we make the transition, we will have a far more secure source for our energy and will eliminate the potential risk of adverse impacts from CO_2 emissions as well. Its source has been with us throughout the ages, as we return to Ra, ancient Egyptian god of the Sun, giver of life.

Let's examine the 5 attributes of sun power before we turn to the technical aspects of solar energy use and its economics as one of our base load forms of energy for powering the United States.

First, it is important to recognize that the answer for the United States and many other countries may not be the right answer for all countries. It's probably apparent that Iceland, for example, will not be basing their future energy plans on direct solar energy conversion. Fortunately, Iceland has tremendous amounts of geothermal and hydro-electric power.

A vivid example of the difference in intelligent energy solutions for countries is in Europe. Austria estimates that by 2010 the country will generate 78% of its electricity through renewables, and Sweden 60%. This is possible because of the high availability of hydro-electric power; coupled with relatively small populations in comparison to their hydro-power resources.

For countries that have the luxury of abundant and consistent sunshine, as occurs in vast tracks of land in the American Southwest, large parts of Africa, South America, Asia and Australia, there is a vast reliable resource.

So the sun is a (1) reliable and (2) safe resource in many climates. Does it have sufficient (3) scale to satisfy our energy needs? As you will see in the following chapters, there is more then sufficient scale to satisfy virtually all energy usage growth far into the foreseeable future; without any new technology discoveries needed. Can it supply the energy needed for electricity, heating, and transportation? As you will see the answer is yes.

Is the solution (4) economic in the midterm and long term? For electricity and heating, you will see in the following chapters, again, the answer is yes. For transportation the answer is yes in the short term for limited markets, and yes in the medium and longer term for most ground transportation markets.

Does the solar electric solution compete with other critical resources? Not only does the solar electric base load solution not compete with other critical resources, it uses resources that are abundant. Uninhabited desert and arid land has little other economic value. Hence, we will be making use of under-utilized assets that would otherwise be unproductive.

The solar electric solution is clean and completely sustainable. In fact, sunlight is arguably the cleanest possible form of energy we will ever use. Also, solar electricity is the most environmentally friendly. Once built, solar power plants have essentially zero pollution and no emissions. Because of the inherently distributed nature of solar energy production, there are no single points of failure and no high value facilities at risk either from accidents or terrorism. The expanded uses of solar electricity will afford investments for increasing the redundancy of the electric grid, which will increase its overall reliability.

The accompanying map shows the locations around the world where solar thermal farms are well suited to provide electricity. These farms will be described in the following chapters. It is apparent from the map that the United States is in the unique position of having superb quality solar lands, vast uninhabited acreage for solar farms, and its population largely in relative proximity to this solar resource.

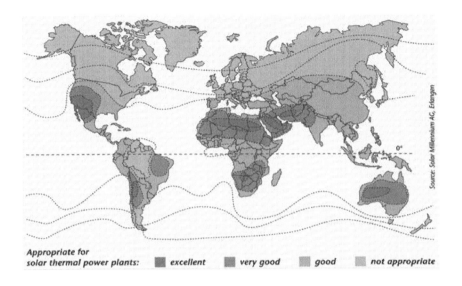

Appropriate for
solar thermal power plants: ■ *excellent*   ■ *very good*   ■ *good*   ■ *not appropriate*

The solar farm acreage needed to fully satisfy the U.S. energy demands for this century is a small fraction of the uninhabited desert acreage available. The vast majority of these lands are unoccupied and of little other economic use. The new technologies and engineering for solar farms makes them highly economical, reliable, safe and scalable.

Nine solar thermal plants built in the California desert from 1985 to 1991 are in operation. New solar thermal farms are being built now in the South and Southwestern U.S. as well. Juno Beach Florida-based FPL Group Inc. runs seven of the nine older solar farms. They have a combined capacity of 354 megawatts, enough to power 230,000 homes.

FPL's solar thermal site in the Mojave Desert, 90 miles northeast of Los Angeles, receives 340 days of full sunshine a year. The parabolic reflectors have an efficiency of more than 90 percent. They need very limited water from 4,000-gallon (15,000-liter) trucks to spray water weekly to clean the reflectors, which are two meters above ground level. With the newest technology and engineering, solar thermal farms can be rapidly scaled to supply the vast amounts of reliable, cost-effective electricity to power our future.

Wind power can play a role in our energy mix for the United States and most countries. But wind power has neither the scale nor reliability of solar as a primary base load power source. The Southwest United States will ultimately be recognized as the Saudi Arabia of solar energy with its massive scale and cheaply available land. The oil fields of Saudi Arabia will have several more decades at best to produce oil for export. America's southwestern desert lands, photovoltaic farms and rooftops around the country and other regions around the world can supply solar energy for billions of years. The technology exists now and is only getting better and cheaper, so many would ask, what's the problem?

The problem is 3-fold. Lack of education about the essential attributes of the required solutions, influence peddling to protect, promote and expand existing industries, and a flaw in the U.S. Constitution. A goal of this book is to illuminate a practical solution for America's power and transportation needs for safe and affordable energy through solar power and shining a light on the entrenched special interest groups that are misdirecting our valuable resources and tax revenue.

How does influence peddling interfere with a solar electric solution? In the case of nuclear power, for example, there are enormous subsidies that distort the real costs of nuclear power and hide the dangers of nuclear power plants and their nuclear waste.

Coal power plants currently pay either nothing or very little for the CO_2 and other pollutants generated from coal, largely through lobbying of elected policy makers and regulators.

Influence peddling and misdirected resources play an even bigger role in our national corn ethanol policy. The U.S. Constitution provides each farm state that grows corn and other biofuels 2 senators, the same as the large population states like California and New York. A state like Iowa has the same representation in the Senate as California. Iowa has approximately 3 million people. California has 37 million people. Having every Californian, New Yorker or Texan who drives a car pay between $0.45 and $1.01 or more per gallon of ethanol for a subsidy to ethanol producers may make farm state senators popular in their home states, but it's very costly to consumers who eat, drive, and pay taxes.

Also, corn ethanol requires almost as much energy to produce as it yields in usable energy. With the competition for corn for ethanol and food, you have all the makings of an energy policy only a political lobbyist could love. Why did the large states' Congressional House Representatives not stop this folly from becoming legislation? Stupidity? Fear of Global Warming? Doing the politically popular thing? Perhaps. It's universally recognized that corn ethanol has already led to many adverse consequences including increasing global hunger and the higher price of fertilizer for all farmers.

The pressure to reduce ethanol, nuclear and implicit coal subsidies will grow, but the forces to maintain them will be even stronger in the short term. It will require a clear path to a better solution and enormous public pressure, and political effort at the local, state and federal level before these enormously expensive policies are scaled back. Fortunately, the better solution is visible on the horizon.

3

Solar Thermal Farms – A Solution for Clean, Economic and Unlimited Energy

The previous chapter laid out prerequisites for solutions for national base load energy supplies. Now we are going to examine the potential role of solar energy in supplying power for electricity, ground transportation, heating and the required enabling solar technologies.

How well does solar energy for the U.S. fit the required attributes as a base load solution for each of the primary electric, transportation, and heating markets? Let's first examine the potential role of sunlight conversion for electric applications such as lighting and industrial power applications.

One of the fallacies voiced over the past 30 years has been the assertion that solar energy and renewables can only play a small part in the overall energy solution mix. This presumption has mainly been driven by cost and availability assumptions, as well as special interests pushing their own energy agendas. What you will find is that there is no fundamental limit for solar energy's role in providing power for the U.S. Oddly, articles and authors who promote solar energy's potential to meet U.S. energy needs sometimes inadvertently create a perception that solar energy is less efficient and less practical as a base load power source then is actually the case.

An example is the January 2008 *Scientific American* article, "A Solar Grand Plan", which described the potential for solar photovoltaic farms to provide base load electricity for the U.S. The article was based on the assumed use of cadmium telluride (CdTe) thin film solar panels deployed in solar farms in the desert Southwest and storing the energy by compressing air in underground caverns for nighttime or cloud covered daytime generation. The underground stored compressed air would drive turbines for electricity generation when the sun wasn't shining.

Cadmium telluride thin film photovoltaic material is approximately 9-10% efficient. There are serious questions about the availability of one of the two main components, tellurium; an extremely rare element in the earth's crust. Because the technology choice for the article was CdTe, the low efficiency assumptions led to solar farm areas about twice the size of what is needed by current solar thermal farm technology. This led to a review in the Wall Street Journal that focused on the land area and inefficiency of CdTe, casting doubt on the ability of solar farms to provide a practical base load backbone for the United States.

With the right choices of existing and near term solar technologies, including overnight heat storing solar thermal farms, there is no practical limitation for solar farms to grow capacity and provide the majority of our 24 hour per day electricity needs nationwide.

Is solar energy conversion to electricity reliable? The conversion technology needs to be completely reliable. We also need highly reliable electricity generation day and night. There are two primary technologies for conversion of sunlight into electricity. The first is photovoltaics (PV), which has been in use for over 40 years and has proven to be reliable. PV panels can last in excess of 25 years with high reliability and little degradation in performance. Many currently sold photovoltaic panels come with a 20 year performance guarantee.

PV cells and panels have proven to be so reliable that they are routinely used in remote applications. These demanding applications include ocean buoys, remote mountaintop microwave relays, satellite power sources for communications, and many other high value applications that demand high long term reliability.

The other major solar electric conversion technology is solar thermal-to-electricity. This technology is nothing more then reflectors concentrating sunlight to heat a fluid, either water, molten salt, or another working fluid or gas, which then drives a turbine to produce electricity.

There are three primary types of solar thermal-to-electricity generation technologies: solar troughs and linear reflectors, solar power towers, and large solar concentrating dishes. Solar troughs and solar power towers are able to efficiently store the thermal heat generated from sunlight for overnight electrical generation.

Solar trough or linear array farms are collections of linear parabolic, curved, or flat reflectors laid out in parallel rows that track the sun and focus the sunlight on vacuum insulated pipes. The pipes are filled with a working medium, such as water, that is heated and converted to steam. These pipes are connected to traditional steam driven electric turbine generators. The superheated steam is generated from the direct solar radiation reflected from the troughs and focused on the thermal pipes and carried either directly to the electric generators or sent to thermal storage tanks for heat extraction. Solar trough farms on a commercial scale have been built starting in the early 1900's in the United States.

In 1910, Frank Shuman founded the Sun Power Company, which designed and built a commercial scale solar trough system for pumping water in Tacony, Pennsylvania, and a commercial demonstration plant in Egypt two years later. Over the ensuing decades numerous refinements have taken place to increase the efficiency, lifetime, and ease of manufacturing the solar troughs or linear flat reflectors, associated pipes, and hardware for power generation.

The design and materials used in today's solar "trough" type systems are highly refined compared to previous generations. Reflector designs have reduced manufacturing costs, increased surface resistance to wearing and scratching, and concentrate sunlight more efficiently. The heat pipes include advanced absorbing materials and are vacuum insulated to minimize heat loss.

Today's leading solar trough manufacturers can achieve in excess of 17% total conversion efficiency from sunlight to electricity and are developing the near term capability to supply and build solar thermal farms measured in square miles per year. Longer term, solar thermal manufactures will have the ability to develop factory capacity to build solar farms covering tens to hundreds of square miles per year.

Leading solar thermal manufacturers Ausra, Solel, Abengoa, and others are developing the manufacturing capability to construct individual solar farms in the desert Southwest and globally, whereby each solar farm can equal the power generation capacity of a coal, natural gas or nuclear power plant.

Ausra's Solar Farm Compact Linear Fresnel Reflectors

It is commonplace in the Midwest to drive past mile after mile of corn, wheat and soybean farms. These vast croplands need to be tilled, planted and harvested every year. It will be equally unremarkable in the future to drive past mile after mile of solar thermal or solar photovoltaic farms, which once installed will only need minimal maintenance services and last for decades.

With servicing or replacement of solar thermal reflectors and moving parts, these solar farms can last longer then 40 years. They require no fossil fuel consuming fertilizers or pesticides, nor fossil fuel consuming transportation costs, once they are built. They are immune to droughts, plagues and pests. Solar farms take up less space then the area that most lakes use for an equal amount of hydro-electric power generation.

Clean power will come from the sun-belt of our earth- the U.S. desert Southwest, northern Africa, Australia and other locations are ideal locations for solar thermal power plants. Large industrial and urban centers that are remote from solar farms can and will be reached with long distance high voltage DC (HVDC) transmission lines. These HVDC lines will be described later in the book.

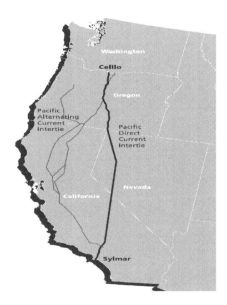

Longest U.S. HVDC line (846 miles/ 3,100 MW capacity) provides electricity from the Northwest to Los Angeles with very low line losses

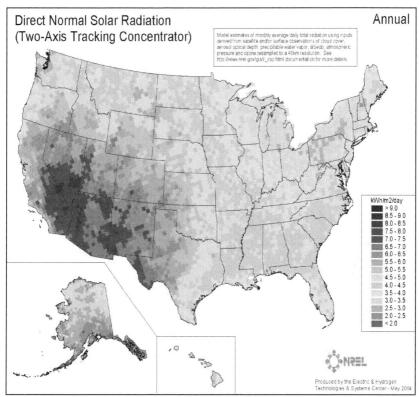

Map of solar electricity generation potential from solar thermal

Solar towers are another type of solar thermal farm that include fields of individually mounted mirrors that track the sun. These farms include a central tower where sunlight is focused and heats a working fluid or material to generate electricity. The absorbed heat can also be stored for electricity generation after sunset or the following day. The first large scale solar tower was built near Barstow, California in 1981 funded by the Department of Energy (DOE). The tower and heliostat reflectors covered 126 acres and generated 10 megawatts. Later it was re-designed to heat molten salt.

A variety of fluids were evaluated by the DOE, including water, air, oil, and sodium, before molten salt was selected. A blended molten salt (a mixture of 60% sodium nitrate and 40%

potassium-nitrate) was used in the solar power tower system demonstration program because it is liquid at atmospheric pressure and it provides an efficient, low-cost medium to store thermal energy. Molten salt's operating temperature range is compatible with today's high-pressure and high-temperature steam turbines, plus it is non-flammable and non-toxic. In addition, molten salt is used in the chemical and metals industries as a heat-transport fluid. Industrial experience with thermal molten-salt systems is relatively well established, but requires thermal storage design engineering for the differing configurations of solar thermal farms.

A Future 4 Solar Power Tower Configuration

A close-up of a BrightSource Energy Power Tower in Israel

Solar thermal electricity generation is best suited for large solar farms in dry, sunny arid regions like the deserts of the southwestern United States, North Africa, Australia and South America where there are long, uninterrupted periods of bright cloud free sunshine.

Because solar thermal farms are based on heating fluids up to 500 degrees C (Celsius) or more, they can store thermal energy overnight, enabling electricity generation 24 hours a day. The fact that storing heat is more then 50 times cheaper then storing the equivalent amount of electric energy in batteries means that solar thermal farms can provide a potentially reliable and economical way of generating electricity day and night without new storage technology breakthroughs.

Demonstration solar thermal farms have been in operation for decades. In 2003, there were nine solar farms generating 354

MW of commercial solar thermal electricity in the United States. As the price of energy has risen and the continued efficiency gains and cost reduction through design innovations has progressed, solar thermal farms are for all practical purposes, nearly economically competitive with natural gas for large scale electric powered generation.

Andasol Molten Salt Solar Thermal Storage Tanks Under Construction Near Granada Spain

The Spanish renewable energy company, Acciona, completed a 64 megawatt solar thermal trough farm near Las Vegas, Nevada in 2007 which supplies power to Nevada Electric and other state utilities.

Nevada Solar One was built with 760 parabolic troughs that concentrate sunlight onto tubes running laterally through the trough focus zone and contain a heat transfer fluid. The concentration ratio is 71:1 (71 suns), meaning the sunlight is concentrated by the reflectors onto the heat pipes by a factor of 71. A special heat transfer fluid flows through the receivers, which achieve a temperature of nearly 400° Celsius. The heated fluid is pumped to the main power plant generator and passes through heat exchangers before generating the steam needed to drive the electricity producing turbines. The troughs track the sun all day and operate from nearly dawn to dusk more then 340 days per year.

Acciona 64 MW Nevada Solar One Farm

Acciona Nevada Solar One Trough Reflector

In 2008, the large California utility Pacific Gas and Electric signed contracts for more then 900 megawatts of solar thermal farm generated electricity effectively at prices that are not much higher than the current cost of new natural gas powered electricity.

Overnight thermal storage for 24 hour electric generation has also been demonstrated, enabling solar farms to grow into 24 hour per day electricity base load generators. A 50 MW solar farm in Spain is one of the newer thermal storage designed solar thermal farms in commercial operation using molten salt for overnight electricity production.

How reliable is sunlight in the desert Southwest for U.S. reliance on solar farms and what is needed for those occasional cloudy days? In the eastern California deserts and parts of western Nevada, the average number of sunny days per year exceeds 345

(94.5% solar availability).   Solar thermal farms can also be designed to include natural gas cogenerated electricity using the same solar farm steam turbines.   This means that including the overnight thermal storage and average solar availability for the deserts of eastern California, Nevada and parts of Arizona, less then 6% of non-solar electricity generation time is required to replace electricity capacity from those solar thermal farms.

Solar thermal farms can cover large areas, making them very distributed and resistant to single source failures, local accidents and terrorism.   Since existing natural gas power plants are easy to maintain and electrical generation capacity can be turned on or off quickly as needed, they can be kept as backup power plants, as many of them are currently.   We will then have a reliable combination of solar farms with thermal storage, legacy nuclear and coal plants, wind power when available, and natural gas power plants that are relatively clean burning. Wind generation capacity can serve to further limit the amount of natural gas power plant generation time needed due to cloudy conditions over the solar farms and solar photovoltaic covered rooftops.

Most areas of the U.S. see peak annual electrical demand during summertime.   Since the highest solar insolence and peak daytime summer electrical usage are highly correlated with hot sunny days, peak solar farm and rooftop PV output occurs during times of highest electrical demand.   This correlation means that as large scale solar base load capacity is built up, it is directly contributing to supplying required peak generation capacity.   This high correlation is important to the positive economics of the initial scale up of solar farm and PV electrical generation capacity.

In the case of wind power, there is little correlation between the peak electrical grid demand in a given region and when peak wind generating capacity can be supplied to the electrical grid, as can be seen in the accompanying graph.

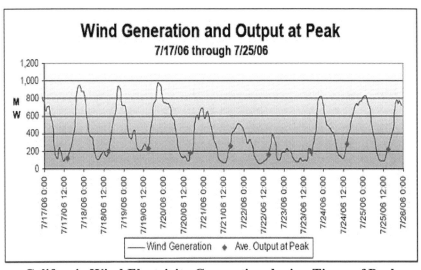

**California Wind Electricity Generation during Times of Peak
Statewide Demand
(Note Poor Correlation in Wind Electricity Supply and Peak
Demand)**

Solar farms can therefore be augmented with wind power, existing nuclear, coal and natural gas power plants to satisfy needed reliability as well as safety attributes needed for 24 hour base load electric generation. That's well and good for residents of the Southwest, but what about the rest of the U.S. and other countries that are not blessed with such abundant sunlight? Fortunately, as you will see, there is not even the need for new solar technology, although new technologies will continue to lower the long term cost of solar energy.

One of the previously mentioned solar concentrating technologies is the concentrating solar Stirling Engine. An example of this technology is the reflective concentrator dish from Stirling Engine Systems which measures 38 feet in diameter and includes 82 curved mirrors. The dishes track the sun and focus its heat onto a Stirling "heat" engine that generates up to 25 kilowatts of electricity. Tests of these "Sun Catcher" systems at Sandia National Laboratories found them to be 31 percent efficient in converting sunlight to electricity. Stirling Engine Systems has signed agreements to supply California utilities with their Stirling Engine concentrators for the generation of hundreds of megawatts of solar electricity.

Stirling Engine concentrators are highly efficient in sunlight conversion to electricity, but do not store thermal heat efficiently for nighttime electrical generation. Thus, Stirling Engine concentrator's utility is best suited for daytime peak power electrical generation capacity. They are also somewhat less sensitive to intermittent cloudy weather conditions then are the larger solar thermal farms.

Concentrating Solar Stirling Engine

The construction of a new high voltage DC (HVDC) grid backbone in the U.S. will allow long range electricity transmission to most parts of the U.S. from power generated in the Southwestern deserts, other regional solar farms and wind farms day and night. The HVDC technology is described in Chapter 7.

One of the main application differences between photovoltaics and solar thermal farms is in their scale. Whereas solar thermal farms require very large land areas to be efficient, PV systems can be effective in smaller scale settings.

It can make perfect sense for a homeowner or business to install PV panels on the roof to reduce the need for utility supplied electricity or sell power into the grid. As the cost of PV panels and installation costs decline, it will become an ever more economical option.

PV systems can also be installed in larger grid-scale farms. As the efficiencies of PV technologies increases and costs decline we will see wider use in home, business and grid-based installations. They also function effectively in partly cloudy or intermittently cloudy climates. They can function effectively in much broader climate regions then solar thermal farms and can take advantage of existing structures, buildings and roofs to lower the cost of installation.

Many U.S. based warehouses with large roof areas in the U.S. have installed PV panels to supplement their own electrical needs, or to sell electricity back to electric utility grids.

PV panels are well suited to direct conversion of sunlight to electricity when the sun is shining. Cloudy conditions significantly limit their electricity production, although some PV technologies such as cadmium telluride are better then silicon panels in such conditions. PV panels do not produce electricity at night which is one reason why solar electricity is often considered a daytime-only power source by many people and media organizations.

Inherently, PV systems are not good for generating power that is easily stored cheaply, unlike solar thermal farms, which can store energy cheaply for overnight electricity generation. This is due to the fact that electricity is much more expensive to store then thermal (heat) energy. Another application difference is that solar thermal farms by their nature are only economic in large grid-scale installations using large high efficiency turbines and larger more complex power grid connections.

PV installations on roofs have zero, or very limited land use impacts in urbanized, suburban, and productive farming areas. Solar thermal farms have large land use needs but only in virtually uninhabited desert regions where land costs are low and the land is generally not otherwise in economically productive use. There are impacts on the local habitat, but there are always trade-offs for land use, even for farmland and areas flooded for hydro-electric dams.

One major efficiency improvement for solar thermal farms that has emerged recently is the reduction in area required for a given amount of electrical generating capacity.

The compact linear Fresnel reflector design developed by the Australian company Ausra, for example, requires only about 3.5 acres per megawatt; about half the area required by more traditional solar trough farms. Ausra's type of reflector is able to achieve this efficiency gain because their compact Fresnel reflectors are flat compared to standard deeply curved reflective troughs. These flat reflectors allow the Fresnel reflectors to be spaced more closely without shading each other at lower sun incident angles. The nearly 50% reduction in land area required for the same solar farm power output will result in large land and cost savings and is an example of how costs continue to fall for newer large scale solar farms.

Solar Farm Aerial View at Kramer Junction, California

Traditional Deep Trough Reflectors
Solar Farm at Kramer Junction, California

A study by Dr. David Mills and Robert Morgan at Ausra, shows the total area of solar farms using a compact Fresnel reflector design such as Ausra's would need to cumulatively cover only 114 miles by 114 miles to supply over 90% of the power required for electric grid <u>and</u> ground transportation needs in the U.S., if all car and truck transportation needs were met by electric vehicles and electricity.

The solar thermal farms could be disbursed across California, Nevada, Arizona, New Mexico, Texas and potentially southern Colorado and Utah. Each state's average area covered could be no more then 44 by 44 miles, if the land areas were divided equally.

Electrical Demand Impact from Plug-in Hybrid Vehicles (PHEV)

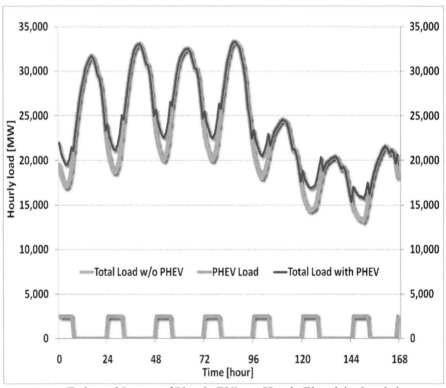

Estimated Impact of Plug-in EV's on Hourly Electricity Loads in Illinois (Argonne National Laboratory Estimate)

Clearly, the land area "scale" for solar thermal farms and photovoltaics to provide base load power for traditional electricity use and EV transportation power is available. How does the area needed for solar farms compare with the U.S. farm acreage planted for corn, which is looked at as a partial renewable substitute for oil as a transportation fuel?

The 2008 estimated U.S. corn planted acreage covers 134,000 square miles. Comparatively, solar farms would require an average of 44 by 44 miles in each of 7 states, covering a total of 13,000 square miles; less then 10% of the total U.S. planted corn acreage. This solar area is for nearly all electric-grid and ground transportation needs for the United States. The chapter on biofuels will show what a special interest farm state boondoggle corn ethanol and biofuels subsidies have become.

If the U.S. were to move to solar farms providing 50% of our electricity and ground transportation energy needs, we will need less then 5% of the corn planted acreage. With roof areas covered by photovoltaics included, the amount of area required by solar farms drops even lower. It now becomes apparent that we have considerable room for energy demand growth, and solar conversion efficiencies will only improve over time.

Wind turbine electric generation capacity is also growing rapidly in the U.S. and may reach between 10 and 20% of U.S. electric generating capacity within the next two decades.

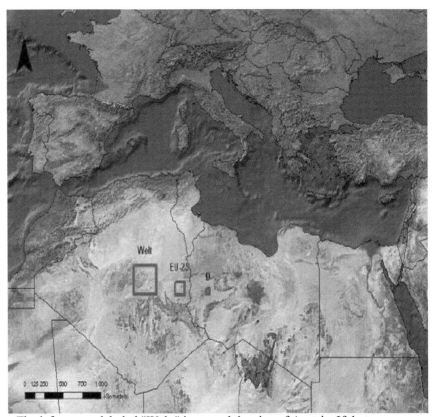

The left square, labeled "Welt," is around the size of Austria. If that area were covered in solar thermal power plants, it could produce enough electricity to meet world demand. The area in the center would be required to meet European demand. The one on the right corresponds to Germany's energy demand.

We now see the potential complementary nature of solar, wind and natural gas generating capacity. Widespread geographic installation of PV systems on existing structures for intermittent daytime and peak load generation, large solar farms in the ideally situated arid and desert regions for 24 hour nearly continuous power generation combined with geographically distributed wind power and existing legacy power generation.

In 2003, a German physicist, Dr. Gerhard Knies, approached the Club of Rome, a global think tank that consists of scientists, businessmen and current and former heads of state from around the world, hoping to enlist their help in determining the feasibility of solar thermal farms to supply Europe with electricity. They were interested, he says, but skeptical, "They said we'd need a disaster first before people will change."

That same year Dr. Knies and other members of the Club of Rome formed the Trans-Mediterranean Renewable Energy Cooperation (TREC). TREC began to further develop their ideas into a more detailed concept, called the Desertec concept. The plan would be to build an array of solar thermal power plants in North Africa and the Middle East that would transmit electrical power to Europe.

TREC concluded through a number of studies that the same types of solar thermal farms that had been demonstrated in the Southwestern deserts of the U.S. were practical for North Africa and southern Europe as well. In April 2006, TREC issued a report by the German Aerospace Center which found that by using less than 0.3 percent of the deserts in North Africa and the Middle East, thermal solar power could meet a major fraction of Europe's electricity needs at highly economic and competitive rates.

TREC presented the Desertec proposal to the European Parliament in Brussels in November 2007. The proposal estimated the costs of the project at $400 billion over 30 years, to include more than a hundred solar farms initially. Eventually the Desertec plan would allow Europe to meet much of its energy demands by 2050 through the development of solar farms with power transmission under the Mediterranean to Europe.

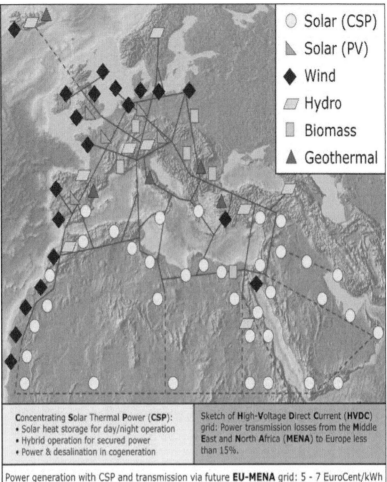

Solar (CSP)
Solar (PV)
Wind
Hydro
Biomass
Geothermal

Concentrating Solar Thermal Power (CSP):	Sketch of High-Voltage Direct Current (HVDC)
• Solar heat storage for day/night operation	grid: Power transmission losses from the Middle
• Hybrid operation for secured power	East and North Africa (MENA) to Europe less
• Power & desalination in cogeneration	than 15%.

Power generation with CSP and transmission via future **EU-MENA** grid: 5 - 7 EuroCent/kWh
Various studies and further information at **www.DESERTEC.org**

The Trans-Mediterranean Renewable Energy Cooperation (TREC) is a network of scientists and politicians who have taken it upon themselves to solve Europe's energy problem. Their vision, which they call Desertec, is to turn desert sunlight into electricity, thereby harnessing inexhaustible, clean and affordable energy.

The DESERTEC Concept of TREC is to boost the generation of electricity and desalinated water by solar thermal power plants and wind turbines in the **M**iddle **E**ast and **N**orth **A**frica (**MENA**) and to transmit the clean electrical power via **H**igh **V**oltage **D**irect **C**urrent (**HVDC**) transmission lines throughout those areas from 2020 (with overall 10-15% transmission losses) to Europe.

Unfortunately, the U.S. does not yet have a coherent national policy to expand solar farms in the desert Southwest and combine them with photovoltaics and a HVDC national electric distribution grid to drive future electric and plug-in hybrid vehicles. The electrification of our transportation system may be the biggest transportation revolution since the development of the automobile.

The growth of the solar power infrastructure in the U.S. will soon begin to reach the tipping point where the economics, reliability, scalability and environmental benefits of solar power begin to become apparent to the public and policy makers. You will be witness to an energy revolution that will likely last the remainder of the 21st century.

4

+ Photovoltaics -
Role in Our Energy Future

The photovoltaic (PV) revolution has accelerated dramatically in the past several years. Increasingly volatile prices for fossil fuels, energy security, man-made CO_2 concerns and technology improvements are at the nexus of a mass manufacturing boom in PV capacity unrivaled in our planet's history. Photovoltaics and solar thermal farms are very complementary to the goal of economic renewable energy for this century.

Photovoltaic solar panels will augment the base load role of the larger solar thermal farms in America and the world as technology improvements and cost reductions drive the PV panel cost per watt towards the cost of natural gas and coal generated electricity. Meanwhile, the cost of coal and natural gas generated electricity has risen and will only rise more over the longer term. This approaching cost parity for grid connected photovoltaic electricity is driving the large increases in photovoltaic production capacity globally.

There are a number of major types of PV technologies in production and under development. Crystalline silicon photovoltaics continue to be the primary solar technology with a greater then 80% market share. The entire PV industry is adding capacity rapidly each year as production volumes increase and demand accelerates for renewable electricity.

Many states are also imposing limits on new coal power plants or reducing the number of coal power plant applications receiving permits. As electrical demand increases and new coal power plants are restricted, electricity prices will rise which will fuel further demand for affordable, clean alternatives.

Silicon photovoltaic manufacturers' ability to rapidly expand production and produce less expensive solar cells has been limited by the cost and supply of solar grade silicon, known as polysilicon. Polysilicon manufacturing capacity is expanding rapidly to satisfy this massive global increase in demand. As the polysilicon supply grows, the PV manufacturers will be less constrained by shortages that have hampered their ability to lower prices and more rapidly expand PV panel production. As the traditional silicon PV manufacturers expand their capacity, advances in thin film PV technologies including amorphous silicon, cadmium telluride, cadmium indium gallium selenide (CIGS), gallium arsenide (GaAs) and other technologies are also rapidly revolutionizing and expanding PV capacity, uses and placement of PV materials and panels.

A new expanding form of PV material is known as BIPV's (building integrated photovoltaics). These BIPV's are beginning to enter into mainstream construction as roofing tiles, sidings, shade coverings, parking covers, and other urban uses. BIPV's are beginning to demonstrate the true power of PV technology to cover large parts of a building's outer surface with power producing coverings while maintaining the building's aesthetics.

Traditional PV panels are also now being used in solar PV farms as smaller versions of their bigger solar thermal farm brethren. PV farms generally are planned to be no larger then 1-10 megawatts, as compared to solar thermal farms which generally are planned in the 200-500 megawatt range.

In terms of efficiency, traditional silicon crystal PV's are reaching 23% cell conversion efficiency of sunlight, and thin film PV technologies are now reaching 10%, as compared to 17-19% for advanced solar thermal farm efficiencies, 31% for solar concentrating stirling engines, and 40% for GaAs PV multi-junction solar concentrators. The U.S. National Renewable Energy Laboratory (NREL) has recently achieved a record 19% efficiency for laboratory produced CIGS solar cells.

During the last several years individual factories and companies have developed gigawatt per year PV manufacturing capacity. While solar thermal farms spread in the United States desert southwest, PV farms, PV panels and building integrated

PV's will increasingly find use in urban and rural settings and as stand alone smaller scale solar PV farms. Currently the cost of traditional silicon PV solar electricity is more then the cost of solar thermal farm generated electricity, with both technologies declining in absolute cost each year. The cost difference between solar thermal farm and PV generated electricity has been declining over the past few years and will likely continue to narrow over the next decade.

Although PV generated electricity is not as cheaply stored as solar thermal farm power, PV generated electricity has many advantages in being integrated on rooftops, building awnings or carports. Another difference between PV panels and solar thermal farms is that PV's are more practical in cloudier geographies where solar thermal farms could not operate continuously or as cost-effectively. The mix of the large base load solar thermal farms and the rooftop or building integrated PV materials will provide geographic diversity, cloudy weather tolerance, overnight generation capacity, and redundancy to a national solar electricity infrastructure.

Rooftop Installed Solar PV Panels (Courtesy Akeena Solar)

A potentially intriguing storage mode for PV generated electricity is hydrogen storage.   The continuing reduction of hydrogen electrolysis (hydrogen from water) production costs for both home and industrial applications may allow for the eventual economic storage of PV electric energy as hydrogen gas for home, transportation or industrial uses.   More about the potential for hydrogen generation for home, industrial and transportation uses of solar electricity will be discussed in the chapter on electric vehicles.

Photovoltaic Background

PV solar cells generate power by producing direct current electricity from sunlight, which can be used to power equipment, recharge batteries, or fed into a traditional electric or HVDC grid. The first practical application of photovoltaics was in 1958 when PV solar cells were placed onboard the Vanguard I satellite to power the spacecraft in earth's orbit.   Soon thereafter the applications for PV generated electricity expanded rapidly for pocket calculators, remote relay telecommunications equipment, ocean buoys, satellites and many other applications. In those days, the value of photovoltaics was exclusively for off-grid applications due to the extremely high price of photovoltaic materials. Today a growing percentage of photovoltaic modules produced are being used for direct grid connected power generation.   For grid connected PV power, inverters are required to convert the DC electricity generated by solar panels to AC current.

Silicon photovoltaic cells are usually packaged underneath glass sheets into panels and sealed to protect them from environmental degradation.   The panels are generally guaranteed for 20-25 years.   A fast growing form of silicon PV's are amorphous silicon solar cells which are now generally manufactured directly on large 5.7 square meter glass sheets and are less expensive per watt then the more efficient crystalline silicon PV panels.

CIGS and other flexible thin film PV materials are being integrated into flexible and curved shapes such as roofing tiles, wall coverings, awning materials and even clothing without the need for rigid glass surface coverings. In the next 10-20 years, it is likely that some buildings in sunny locations will have many of their outside sun facing surfaces covered with these BIPV materials and generate much of the building's daytime electrical needs through these thin film BIPV materials.

The Japan affiliate of Shell Oil, Showa Shell Sekiyu, announced in 2008 that they would be building a 1 gigawatt thin film solar factory based on copper indium diselenide (CIS) PV material. At full production capacity, the Shell factory is expected to achieve significant cost reductions compared to conventional silicon wafer photovoltaics. CIS and CIGS thin film technologies have the potential for both rapid expansion in capacity and lower prices because of the inherent scalability of the technology and the efficient use of thin film material. CIS thin film photovoltaics use approximately 1/200 of the active material compared to crystalline silicon solar cells.

Increasingly larger numbers of photovoltaic panels are now being used for very large commercial rooftop applications, small free standing arrays, and large PV farms for grid connected power supplying utilities for daytime solar electricity.

There are now hundreds of photovoltaic cell manufacturers, panel makers and PV suppliers located around the world. The largest global photovoltaic manufactures include Sharp, Sunpower, Suntech, First Solar, and Q-cells (headquartered in Germany). Each of these manufacturers is building PV production capacity exceeding one gigawatt (peak power) per year. As a point of reference, a large nuclear power plant generates approximately one gigawatt of electrical power (24 hours a day).

12 Megawatt Solar PV Farm in Erlasee, Germany

Spot prices for solar grade polysilicon have exceeded $400 per kilogram due to the enormous increase in demand for solar panels, compared to $25 per kilogram prior to the recent PV manufacturing boom. As the manufacturing capacity for polysilicon rises dramatically over the next several years, polysilicon prices will fall further and the photovoltaic industry's capacity to manufacture solar cells and panels will expand rapidly, exceeding 20 gigawatts per year.

In addition to the increases in production capacity worldwide, sunlight conversion efficiencies for PV's are also rising. Solar thermal reflectors coupled with steam turbines generate electricity conversion efficiencies of ~ 17% or more and have been demonstrated for many years. Eventually commercially sold single crystal "wafer" silicon solar cells are expected to approach 30%. Other more expensive photovoltaic technologies already exceed 35%. Silicon PV's are currently the only mass scalable PV technology in production with sufficient annual manufacturing supply for multi-gigawatt scale capacity.

The reason why silicon has led in the PV market share race is straightforward. Silicon is the second most abundant element in the earth's crust and is easily mined and the process for conversion to purified solar grade silicon and finished solar cells has been well developed over many years. Other PV technologies are growing, some rapidly, but for the foreseeable future silicon will continue to be the leading high volume technology in the photovoltaic world.

The rapidly growing silicon PV thin film technology, where a thin film of silicon material is deposited on a glass substrate generally have about 10% or less total conversion efficiency. The largest semiconductor equipment manufacturer in the world, Applied Materials, now sells complete production line equipment for producing these thin film silicon PV panels, each measuring 5.7m^2 (61 square feet).

Applied Materials Thin Film Silicon 5.7m^2 Production Line

One of the major drawbacks of silicon PV materials is that as temperatures rise, their efficiency declines.   This temperature degradation is just the opposite as compared to solar thermal farm conversion efficiency where higher temperatures yield higher efficiencies.   Hot desert climates therefore have an additional advantage for solar thermal farms over PV farms.

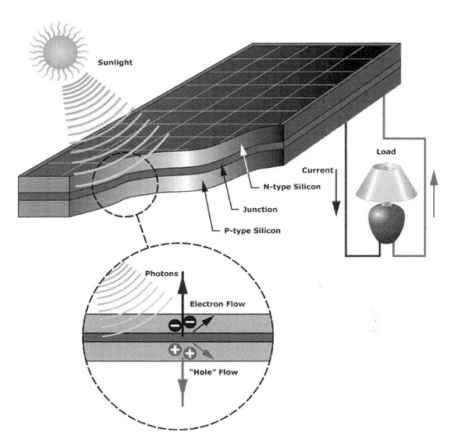

Solar Cells Generate Electricity When Exposed To Sunlight

Will 30% be the upper limit for large scale PV efficiency? New promising technologies make the 30% efficiency number potentially the lower boundary instead of the upper boundary.

Traditional physics teaches that silicon's roughly 33% conversion efficiency limit is due to the material bandgap and the way in which a single electron-hole pair is generated from each photon of light striking the PV semiconductor material. But this assumption is dependant on a single electron being generated from a single photon and the associated conversion inefficiencies. This classical physics description goes back to Albert Einstein's photo-electric effect theory in 1905 for which he received the Noble Prize in physics in 1921. But quantum mechanics offers a potentially more compelling solution.

For conventional semiconductor materials, the one photon generating one electron relationship is the rule. For nano-materials where the material structures have atomic scale dimensions, quantum effects provides a mechanism for the absorption of a single photon and subsequent generation of up to seven electrons which more efficiently capture the photon's energy.

Researchers at the University of California, San Diego, for example, have achieved 45% efficiency for single junction thin film solar cells using nano-particles and quantum wells. It is expected that these cell efficiencies can rise even further as the cell designs are optimized.

From the outside, the UCSD devices behave just like traditional thin-film solar cells. But inside, the nanostructures enable the solar cells to circumvent an important tradeoff that has stymied past attempts to incorporate quantum wells into thin-film solar cells in order to boost device efficiency. Quantum wells can increase solar cell efficiency by raising the photon absorption level and lowering the energy band gap of the PV material.

These nano-material crystals, often referred to as "Quantum Dots", and other "nano" technologies, have the potential to take the solar conversion efficiencies to 50% or higher. Nano-materials also offer the potential to even further reduce the cost of PV panel production by use of thin film production technologies.

Nanosolar, a company funded in part by the founders of Google, Larry Page and Sergey Brin, is based in Silicon Valley and is selling nano-material CIGS PV panels that are manufactured using a roll-to-roll printing method with the cost goal of less than $1/ watt, which would be economically competitive with coal generated electricity in this country.

Future high efficiencies of advanced solar materials will offer consumers substantial opportunity for their own energy independence. Quantum Dot PV panels producing electricity at 60% efficiency would allow a homeowner to generate up to 6 kilowatts of power, or more then 36 kilowatt-hours per day, on sunny days, with panels covering only 10 feet by 10 feet on a rooftop. As a point of reference, the average household in the U.S. in 2006 consumed 30 kilowatt-hours of electricity per day.

As PV panel conversion efficiencies, which now approach 20% commercially, continue to rise towards 40%-60%, rooftop systems will become more and more able to generate a household's daytime electricity from very limited rooftop spaces.

A Nanosolar CIGS Power Sheet Promises Cheap, Flexible Solar Power

PV developmental stage companies like Cyrium Technologies, which is a venture funded Canadian solar PV firm, are already working on modifying PV cells by applying quantum dots. Cyrium Technologies goal of their first-generation PV cells using quantum dots is a conversion efficiency of 41 percent, and a goal for the second generation PV cells is a conversion efficiency of 45%.

One of the truly remarkable future application potentials of these high efficiency thin film PV materials is their integration potential with transportation vehicles. One of the most useful applications will be car integrated thin film photovoltaics. This is where thin film photovoltaics can be integrated into the skin of an electric vehicle. When photovoltaics are integrated into the vehicle roof or exterior, the battery of a sunlit EV will charge so that if the EV is not used for a period of time, the battery will fully charge or have more charge then when it was parked outside. This feature can also ensure that the EV battery does not slowly lose its charge if left for long periods of time.

Some electric car manufacturers have already designed thin film PV's into the car's skin as a way of providing trickle charging power to the car's electronics, so as not to drain the car's primary battery when the car is not running.

As noted earlier, progress has been rapid in the development of thin film nano-technology PV's. If thin film nano-technology PV materials achieve 50% conversion efficiency, then a 1 square meter thin film panel of 50% efficient PV material integrated into the roof of a small EV would charge the EV's battery enough for a trip of 100 miles within six days if left outside in sunny weather.

Two square meters of 50 % efficient PV thin film material integrated into the EV's roof, hood and trunk could allow the EV to travel approximately 20-30 miles per day without any need for external charging in sunny weather. Starting with a charged 200 mile range EV battery, you could drive the EV 60 miles per day for 7 consecutive days without needing to externally charge the battery. The total cost for driving 400 miles if you externally charged your battery at the end of the 7 days would be approximately $7.5 for electricity at 15 cents per kWh of solar generated electricity, or about a dollar per day. The comparable cost with $3 per gallon gasoline would be approximately $40.

Integrated high efficiency thin film photovoltaics would also allow a car to automatically run air conditioning on hot sunny days when the car is parked or unoccupied without running down the car's battery charge.

As the pace of solar thermal farm construction and photovoltaic (PV) capacity accelerates, the bottlenecks for large scale solar electricity projects increasingly will shift to transmission grid expansion approvals and large scale public and private land permitting.

Leasing Rooftops in Order to Accelerate Solar Electricity Expansion

The solar acreage land rush is already in full swing in the desert Southwest. Many investment banks, including Goldman Sachs, have already staked claims and bids on vast tracks of desert BLM (Bureau of Land Management) lands for leasing solar farms. Regulatory and infrastructure bottlenecks will, however, have a significant impact on how rapidly large scale solar farms are developed. Rooftop solar installations, however, have very few regulatory impediments.

A study performed in 2005 by Navigant Consulting estimated residential and commercial rooftop space in the U.S. could accommodate up to 710,000 MW of solar electricity generation capacity if all viable rooftops were fully utilized. Installation and procurement costs are, however, a significant barrier to a building owners' investment in rooftop solar electricity.

New financing options are opening an expanding path to increase rooftop and solar farm installations. Solar power purchase agreements (PPA's), solar leasing and solar financing districts (SFD's) are increasing the options available to property owners.

In many states, cities and counties can create energy finance districts for commercial and residential solar system purchases with loan payments made through 20 year property tax assessments.

The Expanded Federal Solar Investment Tax Credit (ITC)

Tax credits and incentives are still an important component of solar photovoltaic economic viability in the U.S. and most other countries. An expanded federal tax credit for solar electricity was passed by Congress in 2008.

Some of the expanded incentives for solar electric installations including PV systems are:

· Extending for 8 years a 30-percent tax credit for residential and commercial solar systems

· Elimination of the $2,000 incentive limit for residential solar electric installations

· Authorization of $800 million for clean energy bonds (including solar PV)

Greentech Media estimated that PPA's would drive 75% of commercial and industrial solar system sales by 2009, up from 10% in 2005. Residential solar lease financing also has gained traction in the United States.

In order to accelerate solar energy investments, creation of an auction exchange market connecting solar integrators and building owners will also drive market efficiencies and speed adoption of solar rooftop leasing.

Real estate exchange markets have existed for centuries. More recently, auction markets for the buying and selling of CO_2 carbon credits are developing in the U.S. and have existed for several years in Europe. As more buildings are offered up for solar leasing, marketplace liquidity will expand in much the same way as EBay and Alibaba's exchange liquidity have expanded.

Alibaba.com is the world's largest business-to-business (B2B) e-commerce company. It connects millions of online business buyers and suppliers around the world every day.

MMA Renewable Ventures is one of the early innovators of solar PPA's, and in 2008 operated over 40 MW of solar PPA's. MMA's PPA financed projects include the 14 MW Nellis Air

Force Base system, which until recently was the largest PV system in the Unites States, and the 2 MW Denver Airport PV project.

Rooftop leasing and PPA's solve many of the barriers to a building owners' decision for committing rooftop square footage to a photovoltaic system.

Solar integrators have far better technical expertise, purchasing and installation economies of scale then most building owners. Solar integrators are very efficient at capturing incentives from local, state and federal programs. They also generally have better site knowledge for matching cost-effective solar systems to particular settings and climates.

A solar rooftop PPA installation at the University at Buffalo (NY)

Local and state governments are adopting solar electricity PPA's as well. The City of Santa Barbara, California, for example, leased city owned building rooftops for a 330 kW PV system under a 20 year PPA. The project was required to meet the stringent aesthetic standards of Santa Barbara's historic downtown area while providing power to service the equivalent of 1,040 area homes.

The California based company, Solar City, is one of many solar firms that have rapidly grown the leasing approach for homeowners. Solar City has virtually no regulatory delays when leasing solar systems to commercial building or home owners. An investment in Solar City by cadmium telluride thin film PV maker First Solar expanded Solar City's offerings to include First Solar's thin film panels. First Solar panels are well suited for both sunny and cloudier climates and have been sold extensively in the cloud-laden climates of northern Europe. Through solar leasing and PPA arrangements, building owners get guaranteed electricity at lower rates and little or no money down while making productive use of otherwise idle rooftop space. Nearly all performance, reliability, maintenance and financial risks are shifted to the solar integrator.

Rooftop leasing can also significantly increase the number and density of buildings that become electricity generators in urban areas. Many building owners that would otherwise not invest in a rooftop PV system can offer their building rooftops for lease to solar integrators with good pricing transparency from competing vendors.

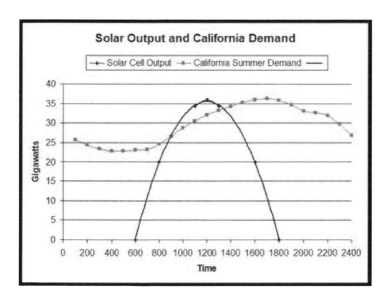

Higher urban solar building densities will make more daytime spare transmission grid capacity available and limit peak summertime power grid overloads.

In 2008, ProLogis, the world's largest owner of warehouse distribution facilities, entered into an agreement to lease roof space to Southern California Edison (SCE), the largest electric utility in California, as a part of the utility's new solar power program.

"This project has the potential to become a breakthrough solar energy program," according to Jeffrey H. Schwartz, the chief executive officer of ProLogis at the time. SCE plans to complete at least 50 megawatts (MW) of solar panel installations every year in its service area. Individual installations are expected to generate one to two megawatts per system.

Solar Installation Costs Continue to Decline

Along with decreasing PV panel costs, installation costs are decreasing rapidly for PV rooftop systems. These cost reductions will serve to increase the value of a building by making the roof PV systems' electricity more competitive and increasing the available profit over the lifetime of the leases.

An example of innovation at work in rooftop installation cost reduction is Solyndra, Inc. For commercial low slope roofs, Solyndra developed thin film CIGS (Copper indium gallium (di)selenide) PV tubular panels that require no roof penetrations, lay completely flat and reduce installation costs by up to 50%.

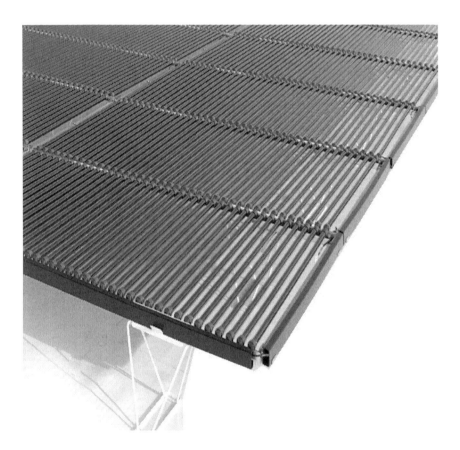

Solyndra Photovoltaic System for Commercial Rooftops

Module installation time for a Solyndra PV system is as little as one day. The panels are safe for flat rooftops even where wind speeds can reach 130 mph due to the panel's unique open tubular design.

The tubular design makes the panels relatively efficient at high solar incidence angles which occur with morning and afternoon sunlight. The panels also efficiently capture sunlight reflected from roof surfaces.

Continued rapid growth of global PV polysilicon supply will require uninterrupted expansion of global PV installations if disruptive oversupply and industry capacity contractions are to be minimized.

Polysilicon prices dropped to $200 per kilogram from $475 in 2008. Jeff Osborne of Thomas Weisel Partners estimated that polysilicon prices would drop to a range of $50 to $80 in 2010 and beyond.

A robust rooftop leasing market and PPA's offer a path for PV integrators to mitigate large solar farm permitting and transmission line infrastructure approval delays. Eventually entire sun exposed commercial building exteriors can be leased for solar electricity generation, thereby generating revenue for building owners, minimizing their PV investment risk and increasing the value of their sun exposed properties.

Ascent Solar is an example of a company developing thin-film (CIGS) building integrated photovoltaics that can sheath building exteriors in photovoltaics using materials like Ascent's thin-film PV aluminum siding. The aluminum siding material is bonded to thin-film solar PV material, providing a building material that is also capable of generating electricity for awnings, shade coverings and even build walls.

As PV panel and installation costs continue to decline, new forms of lease financing and PPA's will contribute significantly to accelerating the shift to solar electricity globally.

5

Sunlight to Electric Vehicles- A New Paradigm

How do we sustain the ever increasing demand for transportation fuel given growing world demand? Do we really need to conserve, drive tiny cars, and drive slowly and less often? Fortunately the answer is no. Just the opposite is true. In this chapter you will see the path linking solar and renewable electricity production with transportation and the technologies that will enable you to drive EV's powered from solar and renewably generated electricity, as well as electricity from any source.

At a lecture in April, 2008 at the University of California, Santa Barbara, a Texas geologist, Jeffrey Brown, presented a compelling case for the decline of exportable global oil supplies and how this would eventually drive the global price of oil higher and force conservation and an end to suburbia. The speaker, being a geologist, was framing the problem correctly. But was his conclusion correct for the longer term? There is actually no shortage of economical, clean energy that can replace oil used for ground transportation. There is simply a transition time and cost needed to bridge the fossil fuel world with the solar and renewable electric transportation world.

If mankind stayed wedded to oil or biofuel based transportation fuel consumption this would be true. Fortunately there is a much better and more efficient energy source for transportation that is consistent with the 5 required energy attributes and the required technologies are already well developed. The integration of solar generated electricity and electric or hybrid-electric vehicles can lead us down the path to economic and environmental freedom from our fossil fuel past.

Rather then cutting back on our transportation desires, we can utilize technologies that already exist and embrace driving powered by pure renewable sunlight. Fuel costs can revert back to the equivalent of when gasoline was less the $1 per gallon. The need for energy conservation to protect the environment or because of high fuel prices may once again be a notion relegated to the past.

Once we link energy generation from sunlight to our transportation fuel source why would we need to conserve? Perhaps for personal economic reasons, but not for environmental, fuel cost or sustainability reasons. In the longer term, electricity, EV batteries, electrically generated hydrogen and EV fuel cell prices will fall as technology improves and the solar infrastructure grows. There is virtually zero pollution generated from sun powered electricity + EV's and we will add virtually no air pollution or CO_2 to the atmosphere once the solar infrastructure is in place.

If you see a large EV Hummer driving down the freeway next to a gasoline Mini-Cooper, and the Hummer displays a "Powered by Sunlight" sticker and the Hummer is made from recycled materials, which car is worse for the environment? Conservation may be appropriate when there is something to conserve. Fortunately we don't need to conserve sunlight.

In the early 1980's, after working on the Voyager Spacecraft Project at the Jet Propulsion Laboratory in Pasadena, California, I founded a company which designed, manufactured and sold high performance lightweight concentrating solar concentrators used primarily as solar stoves sold around the world and even sold at Neiman Markus stores in the U.S.

These lightweight, collapsible solar stoves heat a cooking plate to 325 degrees within a few minutes and cook food and boil water as long as the sun is shining. The devices still cook nearly as effectively today as they did 20 years ago when they were first manufactured. The solar stoves have even been used on expeditions to Mount Everest at base camps up to 19,000 feet for cooking expedition food.

What if you were boiling water with a solar stove to have hot water for tea in the afternoon, and a neighbor walked up to you and said, "if you're not needing the hot water right now, shouldn't you close the solar stove up to conserve energy"? Your reaction would likely be the same as mine. "Why?" It's unnecessary to conserve something that's endlessly available, free and emits ZERO pollution, unless there are other personal reasons to conserve.

With the correct policy choices we will have all the technology we need to begin unlocking sufficient economical energy to supply most of our electricity and transportation needs safely and cleanly. We won't get there safely, cleanly or economically on more foreign oil dependence, food/cellulosic biofuels, coal or even nuclear power.

There is also much misunderstanding about the role of hydrogen in our future energy mix. Hydrogen is a fuel storage medium, the same way a battery is a fuel storage medium. There are no large pre-existing stores of hydrogen to be used as fuel. Hydrogen gas can be created from a number of fossil fuel sources chemically, from biofuels, from electricity or directly from sunlight (although that process is not commercially viable for the foreseeable future).

With the world reaching the limits of increased fossil fuel production, it's unlikely future large scale hydrogen production for transportation fuels would rely solely on fossil fuels, and biofuels grown in the United States are not an economic path for hydrogen production.

Solar electricity can generate hydrogen, as can all sources of electricity, and if EV batteries remain expensive or do not continue to improve their energy density and energy storage capacity at cost competitive prices, hydrogen fueled cars and hybrids may begin to compete economically with electric vehicles or plug-in hybrids.

There are however, ever increasing battery technologies and solutions that may reduce or eliminate the potential need for hydrogen as a future fuel source for EV's. This hydrogen fuel versus electric battery choice will in the end come down to cost and convenience. Each of these two choices offers a completely renewable non-polluting fuel capable of powering our ground transportation vehicle needs.

It is possible that as the price of electricity generated hydrogen and the associated fueling stations declines, hydrogen fueled cars may become an important component of the transportation mix. Hydrogen fueled cars can burn hydrogen in an internal combustion engine, or use onboard fuel cells to generate electricity and operate as an electric car. In both cases, the cars are virtually pollution free, since the only by-product of burning hydrogen is water vapor.

There are numerous hydrogen refueling stations already operating around the world and if hydrogen cars become a significant share of the car market, the infrastructure for hydrogen refueling stations will grow in concert.

There is also increasing interest in providing consumers with home hydrogen electrolyser units for hydrogen generation. With a car converted or designed to run on hydrogen, the homeowner can fuel the car right at home. Hydrogen electrolysers use electricity to generate hydrogen from water, generally through the use of membranes, known as proton exchange membranes (PEM).

The cost of small scale PEM hydrogen electrolysers have been falling, making them potentially practical as a home or industrial source of hydrogen gas. The gas is stored under pressures of between 5,000 and 10,000 psi (pounds per square inch). The pressure tanks can then fill up a hydrogen car or truck storage tank. Cars designed to burn hydrogen are much less efficient in converting hydrogen to useful energy then fuel cells, but these cars are much less expensive and available now as compared to much more expensive and futuristic hydrogen fuel cell cars.

A hydrogen refueling station at Los Angeles International Airport

Now let's look at the enabling technologies for EV batteries that will allow them to be powered by converted solar electricity, or electricity from any source. There are many technologies that can potentially store electric energy and power our transportation vehicles. The focus here will be on the battery technologies that currently are the most mature, have the highest performance, the most economic, and in commercial production or soon to be in commercial production and satisfy our 5 required attributes: Reliable, Safe, Scalable, Economic and No Adverse Consequences.

Lead-acid and Lithium-Ion Electric Vehicle Batteries

The most widely used battery technology for automotive applications is the lead-acid battery. This is the technology that virtually all cars and trucks use to start their engines. It is also the most widely used battery for powering fork-lifts and related factory and warehouse applications. Lead-acid batteries are also the most widely used batteries used in golf cart type electric vehicles, electric bicycles, and the world's current best selling 4 wheeled electric car, the Reva (made in India).

The Reva is a small 2 passenger city car capable of 45 mile range per charge and a top speed of 48 miles per hour. The lead-acid batteries in the Reva can be 80% charged in 2 ½ hours with 100% charged in 8 hours.

REVA "Standard" 2 Passenger Electric Car

The advantages of lead-acid batteries are that they are inexpensive, widely manufactured, and completely recyclable. Lead-acid batteries, however, suffer from several drawbacks. The most limiting drawbacks are their low energy density (kilowatt-hours per kilogram of battery weight), limited number of deep cycle re-charges, and relatively low power density. These shortfalls limit the range of EV's powered by lead-acid batteries and the total miles delivered before the batteries must be replaced. Most lead-acid batteries cannot be deeply discharged more then several hundred times making them less then ideal for high mileage driving applications. There are more advanced lead-acid batteries under development that offer the potential for higher recharge lifetimes and higher power-density. How competitive they ultimately become compared to lithium batteries will be determined in the marketplace over the next several years.

Lithium-ion batteries are the type of batteries that power modern laptop computers, cell phones and many other electronic devices. They are also a battery technology that can offer very high performance for powering electric cars, trucks and buses. They continue to become ever more efficient, safer and more economical. There is little doubt over time lithium-ion and other battery technologies will be better able to store more power with less weight safely, and be much more economical.

Lithium-ion batteries are fast approaching the ability to power EV's with sufficient performance to satisfy public expectations. They will also be able to be powered using renewable solar electricity. As lithium and other battery technology prices drop and newer technologies mature, the cost of transportation "fuel" will drop to where prices were decades ago in inflation adjusted terms. Filling up your car, truck or bus from 100% renewable solar or wind energy will cost no more then the equivalent of $0.50-$0.80 per gallon.

The two main limitations of lithium-ion batteries for use in electric vehicles have been safety and cost. Lithium-ion batteries have safety issues associated with their battery chemistry, whereby the battery can undergo a "thermal event" (catch fire).

The newest versions of lithium-ion batteries designed for electric vehicles have been developed to overcome the risk of a runaway "thermal event", and are generally considered relatively safe for use in electric vehicles. The batteries are designed not only to operate safely if overcharged or overheated, but also if during an accident the casing of the battery is pierced or cracked.

One potential bottleneck in future widespread global lithium battery adoption for EV's is the mining capacity and resource base for lithium carbonate production, which is the material from which lithium-ion batteries are manufactured. Most of the world's lithium production comes from high altitude brine lakes in South America and China. These brine lakes are rich in sodium and potassium salts, and lithium minerals. Solar evaporation ponds are flooded with the lithium rich brines and allowed to evaporate. The concentrated brines are then harvested and the lithium carbonate extracted, purified and refined. The potentially rapid adoption of lithium batteries for EV's will require substantial increases in global lithium carbonate production. It is quite possible that as global oil prices rise and supplies decline, rapid increases in demand for lithium EV batteries will face a bottleneck of lithium carbonate supply until lithium carbonate production can be increased to satisfy this enormous potential demand.

A paper titled "The Trouble with Lithium: Implications of Future PHEV Production for Lithium Demand", by William Tahil, Research Director of Meridian International Research in January, 2007 provides an analysis of the potential issues of increased lithium carbonate supply. It will clearly be a challenge for global lithium producers to meet this potential demand from existing brine lake resources. New lithium carbonate mines will be required to meet this expanding demand even with expanded lithium battery recycling.

It will also be necessary for the EV industry to further accelerate development of alternative battery technologies if lithium production is not able to fully meet the enormous demand for EV batteries as oil depletion rates accelerate in the future.

Other battery technologies are being developed with even higher energy densities and lower potential costs then lithium-ion batteries, but it will be many years before competing battery technologies will be adopted by major car manufacturers. Zinc-air batteries, for example, have exceptionally high energy densities of up to 400 watt-hours per kilogram (Wh/kg), compared to 90-120 Wh/kg for nano-lithium iron phosphate batteries that are currently in production. Zinc air and other battery technologies are, however, much further behind lithium-ion batteries in technology maturity and life cycle time for the large format batteries required for electric vehicles.

Ultimately hydrogen and other types of fuel cell batteries may come into their own as a cost effective transportation technology as their costs come down and their performance rises. Hydrogen's ability to be created directly from renewable electricity at fueling stations or even at home using consumer hydrogen generators makes it a potentially attractive alternative to pure electric battery driven EV's or plug-in hybrids.

Honda Motors and the British company, ITM Power, have developed home and industrial hydrogen generators for use with hydrogen powered cars, in addition to their promotion and development of roadside hydrogen car refueling stations. The Honda home hydrogen unit is fueled by natural gas that is converted to hydrogen, whereas the ITM Power electrolyzer unit uses electricity and water to generate the hydrogen gas. In both cases the hydrogen gas is compressed and stored until needed for refueling a hydrogen fueled car.

ITM's Home Hydrogen Car Electrolyser Unit

Ultimately, long term economics will likely favor pure electric battery EV's due to the very high efficiency of storing and converting electricity to useful motion (turning the wheels). EV batteries are approximately 86% efficient at storing and converting electricity to turn the automobile's electric motors for moving a car or truck. In comparison, electric hydrogen generation is only 50-70% efficient. Also, a car burning hydrogen is approximately 25-30% efficient whereas a hydrogen fuel cell car is 60-75% efficient.

It is largely a question of industry's ability to supply global demand for high performance but affordable lithium-ion batteries that will determine if hydrogen has a near term role in bridging the renewable energy needs of the U.S. in the next decade. Ultimately pure battery powered EV's or plug-in hybrid cars will likely be more cost competitive to operate due to an EV battery's unrivaled storage efficiency and existing electrical transmission infrastructure.

Hoping to influence consumer behavior, some utilities have already created special battery charging rates for plug-in hybrid and pure electric cars. Sempra Energy's subsidiary, San Diego Gas & Electric Co., for example, has created a nighttime electricity usage rate for electric cars that is half that of its daytime rate. If more utilities expand the nighttime reduced electricity rates for residential electricity, the equivalent cost to re-charge your EV may be less then $0.50 per equivalent gallon. A 400 mile trip in a full size EV car could cost as little as $7-10 in electricity.

Lithium-ion and lithium iron phosphate batteries have undergone years of development and testing for use as primary transportation batteries. A more recent development in lithium battery chemistry utilizes nano-particle technology to vastly increase the battery's lifetime and recharge rate. This is a very significant advance for making EV's more economical and convenient. This innovation is used in A123 System's commercially sold batteries for power tools utilizing lithium iron phosphate and Altair Nanotech's lithium titanate nano-batteries. These batteries can provide between 4,000 and 10,000 complete discharges and recharges, or over 500,000 miles of driving. These nano-batteries can also be charged in as little as 10 minutes or less at fast charging stations. The batteries are also considered relatively safe and not prone to overheating and fires that afflicted earlier lithium battery technologies used in laptops and cell phones.

Chile's Atacama Desert currently produces the largest market share of the world's lithium carbonate, which is processed into the lithium used to make advanced batteries, as well as other products including medicines. The brine lakes of this remote desert region are the lithium equivalent to the massive Ghawar oil fields in Saudi Arabia.

Millions of A123 Systems nano-technology fast charging batteries have already been sold in DeWALT power tools demonstrating the commercial success and safety of nano-technology lithium batteries.

As the price of these and other EV battery technologies declines, and improvements are made in energy storage density and new materials, we can expect safe batteries that will power our transportation vehicles either as fully electric cars or as plug-in hybrids. Plug-in hybrids will have a place in the market until prices of EV batteries decline to where the internal combustion engine part of the hybrid engine is unnecessary.

Price declines and energy density improvements in these EV batteries will allow affordable EV's to achieve driving distances on a single charge that approach the range of internal combustion engines. Eventually EV's may even exceed the range of gasoline or diesel vehicles on a single charge.

You will be able to drive further for less money, pollution and CO_2 emissions-free, and have fewer repairs and lower maintenance bills.

Electric vehicles have no need for engine air filters, spark plugs, catalytic converters, far fewer fan belts, fewer moving parts and fewer parts needing routine maintenance. In addition, there is no need to change and dispose of the car's motor oil.

Conservation will stop being about the environment or conserving a depleting resource. You'll be guilt-free to consume as much sunlight generated electricity as you like. As the power of entrepreneurial innovation accelerates, solar and wind generated electricity and EV battery costs should continue to decline and once again make traveling long distances more affordable with larger vehicle choices in competitive price ranges.

Electric Vehicle Choices

The main EV purchase limitations and trade-offs that you will face in the next few years will be the number of choices for plug-in hybrids and EV's, the relative expense of longer range and larger EV's, and the need to recharge them more frequently until battery prices decline further.

For many individuals and families, an EV can be a second car or transportation vehicle best suited to commuting and local driving around town. If you are willing to spend more money for an EV, then you can already purchase high performance sports car EV's such as the Tesla. The Tesla is capable of 0-60 mph in 3.9 seconds and has a 220 mile range per battery charge. The cost per mile for electricity even for the high performance Tesla is only 2 cents per mile.

It is important to understand the choices for EV's that are and will be available to you. There are choices for 2, 3 and 4-wheeled electric vehicles today and the choices will only grow larger in the next several years.

In China, there are hundreds of different electric bicycle models and about 25 million electric bicycles sold there every year. There are also rising numbers of electric scooters on China's roads each year. It is remarkable that China's dominate form of machine powered transportation is already electric vehicles, since there were only approximately 8 million new cars sold in China in 2008. The number of electric bicycles and electric scooters sold in China each year is rising faster then the number of gasoline or diesel cars, trucks and motor scooters.

In many Chinese cities, electric bicycles and electric scooters outnumber passenger automobiles by more then ten to one. Rush hour traffic in many large Chinese cities is completely dominated by electric bikes and scooters.

A City Taxi Brand Electric Bicycle from China with a 36V 10Ah Battery

Most of the electric bicycles in China are powered by lead-acid batteries. The lighter, longer lasting, longer range but more expensive lithium batteries are gaining ground in China and eventually will replace many of the lead-acid batteries. Since most electricity in China is produced with coal, it's clear that China's electric bicycle movement is not yet driven by the desire for a cleaner environment, but rather pure economics and convenience.

Even in this country, you will find there are many choices for electric bicycles and scooters that allow you to pick either a lead-acid or lithium-ion battery pack.

Your path to pure electric and solar electric transportation may start first with the choice of electric bicycles and scooters. For short commutes to work or around town usage, you will find that an electric bike or scooter with a basket or rack can also be very convenient for local shopping and will cost less then 1 cent per mile in electricity. If you ride your electric bike to work and re-charge it at work, then the recharge cost to you can even be free. Parking is free and many towns and communities have dedicated bike paths.

The majority of electric bicycle motors are between 190 watts and 250 watts (1 horsepower is 746 watts). Recharge your electric bike with 15 cent per kilowatt-hour solar electricity and you can ride on an electric bicycle for 4 hours for less then a total of 15 cents.

Electric bicycles also allow you to pedal on your own, so you can choose between either electric only or power-assisted pedal mode. In bicycles with a full electric mode, you can rely completely on the battery without pedaling. Many electric bicycles are capable of taking you well over 20 miles without the need to pedal at all.

You also can buy folding electric bikes that have 16 or 20 inch wheels that allow you more freedom to take your e-bike on public transportation, up stairs, or store it out of the way in your home or office.

50 Mile range, 20-35 mph Electric Bicycle Made by Optibike (U.S.)

Electric scooters are also rapidly gaining ground as a form of clean transportation. A typical electric scooter (a Zapino from ZAP! for example) having a 2.4 kilowatt-hour lithium battery and range of 65 miles, would cost about 36 cents to fully recharge, and it would be completely pollution free and very quiet. In comparison, a gasoline Vespa is polluting and noisy and gets less then 72 miles per gallon, which is more then 10 times the cost of re-charging an e-scooter driven the same distance.

Electric scooters have been illegal in some regions of China. Chinese companies now frequently build electric scooters with the minimum human pedal capabilities to satisfy the legal requirements of an electric bicycle. The Chinese buyers then remove the pedals after purchase and drive the e-scooter without any pedal power hardware attached.

Zapino Electric Scooter from ZAP! Corporation

E-scooters and e-bikes powered by nano-technology batteries will have the ability to be re-charged in minutes with over 2000 complete deep discharges and re-charges with very little performance loss.

U.S. federal law states that electric bicycles with fully functioning pedals, no more than 750 watts of motor power output, and a top speed of 20 mph on motor power only, are to be treated as "bicycles", and are not subject to motorized vehicle laws.

There are also a range of three wheeled electric vehicles that are available to buy. Three wheeled vehicles fall under the safety and regulatory guidelines of motorcycles in most states and are generally inexpensive to buy. Some companies sell three wheeled EV trucks and sedans and have already developed networks of dealerships in the U.S. and Canada.

Many electric vehicle manufacturers are developing additional three wheeled EV's that will be entering the market place over the next few years. Some of the models will be high performance vehicles.

Electric Bicycle Regulations

The Consumer Product Safety Commission (CPSC) has the responsibility for governing the safety of new production model electric bicycles. The CPSC now defines a "bicycle" to include an electric bike with pedals capable of propelling the bicycle with an electric motor of no more than 750 watts, and a top speed using the motor power of no more then 20 mph.

Electric bicycles that fall under this category are not required to be registered or licensed, and no driver's license is required to ride them. They are subject to all the rules of the road, and additional laws governing the operation and safety of electric bicycles may be extended by state or local governments. Electric bike riders have the freedom of being able to use an electric bicycle on public roads and bike trails in every state.

You should check any additional state regulations covering e-bikes, e-scooters, and electric motorcycles in the state you reside if you ride or intend to ride an electric 2 or 3 wheeled vehicle.

California has passed regulations very friendly to electric bicycles, electric scooters and electric assisted 2 and 3 wheeled vehicles.

Texas law provides that electric scooters may be ridden on public thoroughfares provided that the speed limit does not exceed 35 mph. No license is required and the scooters must follow the same regulations as those that apply to bicycles. Counties and municipalities in Texas are allowed to pass there own regulations.

New York's electric scooter laws are some of the most restrictive in the country and severely limit where electric-motor assisted vehicles can be operated. Most state regulations covering electric bicycles and electric scooters fall somewhere between California's lenient and New York's more restrictive regulations.

In California, the law states a "motorized bicycle" or "moped" is any two-wheeled or three-wheeled device having fully operative pedals for propulsion by human power, or having no pedals if powered solely by electrical energy, and an automatic transmission and a motor which produces less than 2 gross brake horsepower and is capable of propelling the device at a maximum speed of not more than 30 miles per hour on level ground.

The following is a summary of California's Electric "Moped" Law which went into effect January 1, 2000:

In California an electric "moped" is NOT required to have pedals, and can not be capable of traveling at speeds above 30 mph on level ground on the motor's power only. Mopeds require obtaining a one time license plate and registration card at the DMV. Moped licensing in California is only $6.00 and never expires and never needs to be re-registered. Drivers must obey all rules of the road, wear a DOT (Dept. of Transportation) approved motorcycle helmet, and be over 16 years old with a valid learner's permit or drivers license. No insurance is required.

California 21220.5 Electric Scooter Law Summary:
- A driver's license is not required to drive an electric scooter.
- No insurance, registration or license plates are required.
- Driver shall not operate a motorized scooter if under 16 years of age.
- Driver must wear a bicycle helmet regardless of age.
- Driver can be arrested for driving while intoxicated on a motorized scooter.
- Drive shall not operate motorized scooter in excesses of 15 mph.
- Driver shall not operate on public road with a speed limit of 25 mph or more, unless it is operated in a bike lane.
- Driver to operate on right side of the roadway and next to the curb when feasible. When intending to move left at an intersection, they must walk their scooter in a crosswalk when crossing roadway.
- Driver must operate in a designated bike lane when one is available.
- Driver must not ride on any sidewalk except to leave or enter adjacent property.
- Driver shall not leave scooter on path or sidewalk.
- Driver shall not hitch scooter onto another vehicle in motion.
- Driver shall not have passengers.
- Driver can be on bike path or trail unless prohibited by local ordinance.
- Driver must have at least one hand on handlebars.
- Equipment requirements during hours of darkness:
  - White headlight to the front visible from 300 feet both to the front and sides.

- **A red reflector on the rear of the device, visible from 500 feet.**
- **White or yellow reflector on each side visible from the front and rear of the device from 200 feet.**
- **A white lamp or white lamp combination, attached to the operator and visible from 300 feet in front and from the sides of the motorized scooter.**
- **Handlebars must not exceed shoulders of the rider.**
- **Scooter must have an engine/motor kill switch that activates when released or when the brakes are applied.**
- **Scooter must have a working brake.**

Electric Vehicle Advantages

The rapid explosion of hybrid, plug-in hybrid, and pure electric 4 wheeled vehicles is symptomatic of the advantages of electric vehicle technology. As lithium battery technology matures and costs decline, the choices for completely electric vehicles and plug-in hybrids will expanding rapidly. The advantages of EV's are many-fold and pervasive.

Whereas existing internal combustion engines are ~ 17%-25% efficient at converting the energy in gasoline or diesel to motion, electric vehicle motors are generally over 85% efficient. Electric vehicles are therefore 4-6 times more efficient in converting energy to useful motion then internal combustion vehicles. Electric vehicles are non-polluting. EV's can be recharged at home and can eliminate an entire supply chain required to transport liquid fuel great distances by truck, railcars, pipelines or ships. Electric vehicle batteries are recyclable and there is no required disposal of spent motor oil, air filters, spark plugs or other internal combustion engine parts. The very long life nano batteries will even be able to be moved from older EV's to a replacement EV at low cost and can continue to be used for additional hundreds of thousands of miles.

Utilities are looking at power storage options based upon used EV lithium batteries for grid stabilization applications to lower their battery acquisition costs.

What are the practical technology and economic cost limits to solar electric power and transportation battery technology? How fast can we get there? The largest opportunities for cost reduction improvements in the solar-to-electric car supply chain are for photovoltaics and EV batteries.

EV Fuel Cost Advantages

The same cost advantages for recharging electric cars holds true for electric trucks, vans, forklifts and buses. The advantages and cost savings are realized most rapidly in utility vehicles such as taxis, trucks and buses that are driven constantly since the cost of EV fuel (electricity) is so much less then gasoline or diesel.

2008 Tesla Electric Vehicle with 53 kW·h Li-ion battery
(0-60 in 3.9 seconds)

The demand for lithium-ion and other EV batteries in the U.S. will expand rapidly as the price of oil based fuels for internal combustion engines rises. Because ethanol in the U.S. is so inefficient to produce and remains an air polluting fuel, it is unlikely that ethanol over the long term will be nearly as clean, efficient or cost-effective as EV's as battery technologies improve and costs decrease. It is also clear that special interests, farm state Senators and taxpayer subsidies will remain the mother's milk of U.S. produced ethanol for the foreseeable future.

Can EV batteries achieve the cost-effective scale and volume required to be a fundamental transportation platform? The only material that might be of concern within the current suite of EV battery technologies is lithium carbonate, from which the lithium in lithium batteries is derived. Some estimates show that there are large available mineable resources of lithium for electric car batteries. However, current lithium production is almost exclusively sourced from brine lakes containing lithium and associated large solar drying ponds in Latin America and China. It does appear potentially problematic whether there will be sufficient lithium carbonate production increases in the short term to handle a looming large scale-up in lithium battery demand. It is likely that lithium EV batteries will provide the initial phase of EV advanced batteries, but to fully convert the majority of the world's transportation infrastructure to EV's other battery technologies may need to be advanced to the same level of car integration as lithium batteries.

New more advanced battery technology improvements are under development for lithium nano-technology batteries as well as other battery technologies that will offer even greater driving range and lower cost. The renewable electricity fueling cost is already quite low and can decline in inflation adjusted dollars as ever better solar conversion technology improves and other cost competitive renewables such as wind generated electricity expand.

Let's examine the cost of fuel delivery to an EV. Charging stations built for EV's will be much more common then gas stations and can even be built into parking lots for the workplace, retail outlets, roadside stops and parking garages. Slow charging outlets cost little more then standard electrical outlets.

Rapid charge stations will need to support high power and high current capacities and will require specially tailored charging equipment that is already currently commercially available. Costs for high capacity quick charge stations would be less then current construction costs for new gas stations and maintenance costs would be lower as well.

Car Charging Station in San Jose, California

The most convenient location to recharge your EV will be at home. For slow EV re-charging overnight where a standard 120V or 220V garage outlet is used, all the electrical needs of the EV are already built into the average garage or carport. How much does it cost to move the fuel (electricity) from solar farms or rooftop PV panels to the car via the electric grid? The delivery costs would be very competitive since consumer owned PV panels would already be part of a house electrical connection and any costs for new electric grid infrastructure build-out would be amortized over each kilowatt-hour delivered over the life of the HVDC electric grid upgrade.

Once the solar farms and HVDC grid reach sufficient scale, the incremental cost of renewable power delivery to a home fueled EV is virtually nothing. Compare the cost to deliver biofuels to a vehicle using corn-based ethanol or cellulosic ethanol. Fuel needs to be expended each season for plowing, planting, spraying, and fertilizing the crops. Then fuel is expended to harvest the crops and move them to a production facility along with the other inputs and fuel needed to used to process the crops. Once the ethanol is produced, it must be transferred to tankers via train or truck. Pipelines cannot transport ethanol due to ethanol's incompatibility with traditional pipelines.

Ethanol is then shipped, often over great distance, to distribution centers where it is either blended with gasoline or other fuels and then trucked to retail outlets. Compare this hazardous and costly supply chain to flipping a switch in your garage and the zero pollution electric fuel flows from solar farms over the grid or directly from your PV roof or BIPV's straight into the EV's battery. Think of all the supply chain trucks for the production of ethanol and fuel trucks taken off the highway and all the fumes, exhaust and the occasionally exploding fuel truck highway accidents that would be eliminated.

California currently has a program to provide rebates for qualified electric car purchases. For example, qualifying electric city cars and pickups are given as much as $1000 cash rebates by California's Air Resources Board as "zero emission vehicles" under a state incentive program. Whereas a traditional driver might fill up the gas tank for about $40-$60 dollars, a small EV can be recharge for as little as $2.00 with off-peak rates. The rebate would be the equivalent of over 30,000 miles of driving assuming electricity costs of three cents per mile. That would mean several years of not having to pay for fuel for a small EV, as compared to a biofueled car where the fuel cost would be about $3,500 for the same 30,000 miles of driving.

Numerous EV's are coming to market, including the stylish Norwegian based Think car which allows several battery options including lithium nano-technology batteries. There are European, U.S., Chinese based EV's and plug-in hybrid cars in development and will be commercially available in ever increasing numbers.

The Th!nk Ox EV

The Th!nk Ox EV may be released in 2011 and is an example of a well engineered, safe 4 wheel, 5 passenger EV with a range of 125-150 miles per charge and several battery options including lithium iron phosphate batteries that are capable of being charged in minutes.   The OX is planning to integrate roof solar panels to power the onboard car electronics. Th!nk cars are also 95% recyclable. The headquarters and the manufacturing plant are based in Norway. Norwegian gas prices typically exceed $8 per gallon.

Ray Lane, a Kleiner Perkins Managing Partner and Chairman of TH!NK North America, said at a Fortune Brainstorm Green Conference held in 2008 in Pasadena, California, "The transportation industry is undergoing its largest transformation since Henry Ford built the model T."

Plug It In or Fill It Up?

Plug-in hybrid vehicles represent a near term mass market electric vehicle that combines the ability to drive great distances without recharging every fifty to two hundred miles. Plug-in hybrid electric vehicles (PHEV's) are a hybrid vehicle with batteries that can be recharged from an electric power source or the vehicle's engine. PHEV's combine characteristics of a hybrid electric vehicle and pure electric vehicle by combining an internal combustion engine and larger batteries for electric-only driving of up to 40 miles.

During typical daily commuter driving, a PHEV's power comes from the battery. A PHEV driver can drive to and from work on all-electric power, plugging in the vehicle at home to charge it at night, and be ready for another all-electric commute in the morning. A PHEV can also be plugged in at work depending on the availability of chargers at the worksite. Employers are beginning to offer parking spaces that have electric car charging spaces set aside for employees. For longer road trips, the PHEV battery can be used until its charge is depleted and then the vehicle switches automatically to the gas or diesel burning engine. In the case of some PHEV's, such as the Chevy Volt, the car will always run off of the battery even while the battery is being charged by the gasoline engine. The car can then have either its gas tank refilled, or the electric battery recharged or both.

PHEV's will be available as passenger cars, vans, utility trucks, school buses, motorcycles, E-scooters, and other transportation vehicles.

There are numerous plug-in hybrid vehicles that have been announced from General Motors, Ford, Toyota, Chinese automaker BYD Auto, and Fisker Automotive. There are also kits and services available to convert standard hybrid vehicles like the Toyota Prius to plug-in hybrids. The main component of the upgrade is a larger battery to allow a hybrid car to drive up to 40 miles on the battery alone. Most standard hybrid cars have batteries only large enough to drive one to four miles exclusively on the battery.

The movement to more fuel efficient cars and higher gas mileage electric hybrid cars is primarily driven by the price of gas and concern over the environment. As the world fully experiences peak oil and long-term expanding Asian demand, the higher cost of oil and gas causes what is know as "demand destruction" from higher oil prices.

How a Plug-In Hybrid Works

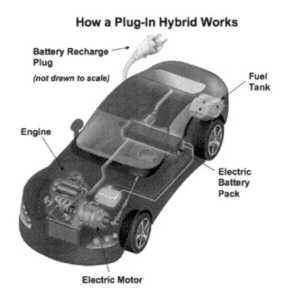

Schematic of a Plug-in Hybrid Vehicle (Source: National Renewable Energy Laboratory)

China's BYD Automotive Corp.'s plug-in hybrid electric vehicle, the F3DM, released in China in 2008, was the first mass-produced plug-in hybrid car to go on sale to the public. The F3DM, initially retailed in China for $21,200 and travels 100 km (63 miles) on its battery before needing to be recharged. The car can be plugged into any 220-volt wall outlet for recharging.

BYD started as a cell phone battery maker in China. The company grew to become China's largest domestic lithium cell phone battery maker. BYD subsequently became a Chinese car manufacturer through a Chinese car maker's acquisition, and then developing its own lithium iron phosphate battery technology for use in its own electric cars.

China's BYD 2009 F3DM Plug-in Electric Hybrid Sedan

Hybrid electric cars that are not plug-in capable only accomplish what higher fuel efficiency vehicles accomplish. They do not break the link between transportation and fossil fuel consumption. Plug-in hybrids and pure electric vehicles begin to break this link. When the electricity supplied is entirely from solar and other renewable energy sources, this link is broken and transportation costs will no longer reflect the increasing limits of fossil fuel production and consumption. The link is also broken for electricity and transportation generated CO_2 emissions into the atmosphere.

There is also increasing focus on the development of heavy duty electric trucks. For example, in California the Port of Los Angeles and South Coast Air Quality Management District, which is responsible for regulating air quality in the Los Angeles area, have co-funded a demonstration of one of the world's most powerful electric trucks developed by Balqon Corp.

The trucks are 100% electric and emissions-free and designed for short-haul operations. The trucks are capable of pulling a 60,000 pound cargo container at 40 mph. They have a range between 30 to 60 miles per battery charge. The battery chargers for these trucks can charge up to four electric trucks simultaneously in four hours and can also provide up to 60 percent of a full charge in one hour to meet peak demands during daily operations.

The Balqon trucks cost about 20 cents a mile to operate, which is four to nine times cheaper then a comparable diesel truck, depending on fuel prices and the amount of truck idling time. Idling diesel trucks are a significant contributor to air pollution as well as wasteful of fuel. In the case of the Port of Los Angeles, there are more than 2 million truck trips per year between the port terminals and rail yards in the port area with much of the trucking time spent idling.

Over time, most fossil fueled ground transportation vehicle classes will be able to be converted to electric or hydrogen powered transportation with significantly lowered operating costs.

The Santa Barbara California Air Pollution Control District Executive Director Terry Dressler highlighted in an interview that electrifying diesel trucks and diesel heavy lift vehicles would result in large improvements in the air quality of many major urban areas. He indicated this was especially true for U.S. cities like Los Angeles and Denver. International cities including Mexico City, Beijing, Bangkok and Jakarta also have some of the worst air quality problems in the world, caused largely by gasoline and diesel cars and trucks.

There Is No Transportation Energy Shortage

Since there is no shortage of inexpensive flat, arid, sun drenched uninhabited land (or rooftops) in the U.S. and no shortage of construction materials for solar farms, there need not be a shortage of clean transportation energy.

The transportation electrical energy costs will follow a cost curve downward as technology improvements delivery more solar energy and more battery energy density and storage capability at lower and lower prices. Because solar energy generation has no major long term hidden environmental or societal costs, the true costs of solar energy generation can fall as supply is increased without land or other resource constraint. Over time the infrastructure for EV battery production and energy density improvements will also lead to declining battery weight and battery costs per kilowatt-hour of storage capacity.

Unfortunately, by far the largest subsidies for "renewable" fueled cars are for food crop ethanol fuel in the United States. There is a need to move more of these incentives to more resource efficient sunlight generated electricity and EV's. As the EV and solar infrastructure grows, we will rapidly see the increasing benefits of solar electricity and the solar electric + EV combination.

6

Day and Night Solar Energy Storage

One critically needed component for the pervasive adoption of solar farms and 24/7 solar base load electric power is the ability to handle cloudy weather and overnight electricity generation. Solar thermal storage technology was demonstrated as early as 1912 in New York using solar thermal heat sinks. The cost of heat energy storage at present is approximately 50-100 times cheaper then battery storage for large scale operations. Adding overnight and cloudy weather heat storage systems increases a solar farm's generation capability to 24/7, but also can increase the size of a solar farm in order to produce electricity 24 hours a day. The additional capacity is needed since during the day, some of the thermal energy is needed to store heat for nighttime or cloudy weather electric generation.

Building the capability for overnight thermal storage combined with the grid infrastructure upgrades needed for efficient transcontinental electricity transmission will provide the U.S. with day and nighttime electricity not only in the southwestern regions, but nationwide as well.

Another beneficial aspect of solar thermal farm heat storage is the synergy with wind powered electricity. When the increasingly large wind powered generation capacity in Texas and the Great Plains is operating at near peak capacity, surplus solar farm heat can be stored for use later that evening or the next day.

Excess solar and wind generated electricity could ultimately be used economically from the grid for hydrogen gas generation through water electrolysis. An incentive for additional hydrogen gas generation could occur through real-time electricity pricing, as currently occurs with wholesale electricity rates.

Thermal heat storage that would be used for solar farms is a relatively well understood technology. One of the most economic methods for solar thermal heat storage is through the use of molten salt. There are also a number of other thermal storage systems that have the potential to provide cost-effective thermal overnight storage of heat that is generated by solar farms.

The principles involved in thermal storage are relatively straightforward. The two main heat storage systems are (1) direct heating of the working fluid or material, or (2) through use of a heat exchange system, where the heated fluid or gas is run through a heat exchanger. The heat exchanger transfers heat to a storage material for thermal storage. The thermal storage system then feeds back the stored heat to drive the electric turbine generators during cloudy periods and overnight. Thermal storage systems have been built and operated that are capable of storing usable heat for days. Thermal storage systems for solar farms can easily be designed to provide stored power (heat) for the turbines for 12-24 hours or longer with acceptable heat loss.

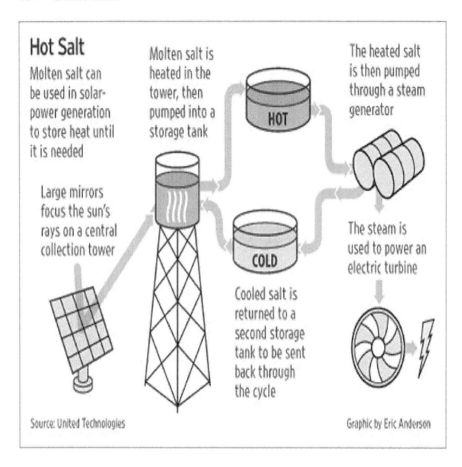

Hot Salt

Molten salt can be used in solar-power generation to store heat until it is needed

Large mirrors focus the sun's rays on a central collection tower

Molten salt is heated in the tower, then pumped into a storage tank

HOT

COLD

Cooled salt is returned to a second storage tank to be sent back through the cycle

The heated salt is then pumped through a steam generator

The steam is used to power an electric turbine

Source: United Technologies

Graphic by Eric Anderson

In previous U.S. Government demonstration programs that operated for many years, the molten salt used was a mixture of 60 percent sodium nitrate and 40 percent potassium nitrate. This salt mixture melts at 430 F and is kept liquid at 550 F in insulated storage tanks. The salt is them pumped to the top of a solar tower, or to a solar trough heat exchanger.

In the case of a solar tower, the receiver at the top of the tower is a series of thin-walled stainless steel tubes. The heated salt then flows back down to a second insulated hot storage tank. The size of this tank depends on the requirements of the solar farm; tanks can be designed with enough capacity to power an electric turbine overnight or longer. When electricity is needed from the thermal storage tanks, the hot salt is pumped to the steam-generating system to produce superheated steam for a turbine generator.

As an example of the size of a molten salt storage tank, tanks that provide enough thermal storage to power a 100-megawatt turbine for 16 hours would be about 30 feet tall and 160 feet in diameter. This would correspond to a 350 acre solar thermal farm using the best current commercial technology. Studies by Sandia National Laboratories and the National Solar Thermal Test Facility have shown that a two-tank storage system can have storage efficiencies of 95%.

Molten salt technology for solar power applications has been evaluated for over a decade by the U.S. Department of Energy and is now considered mature enough to begin use in modern commercial solar farm generating plants. The Oak Ridge National Laboratory successfully built and operated two nuclear reactors which utilized molten salt as the heat transfer medium, one in 1954, and one in 1960. The technology is also considered relatively well established for industrial scale thermal heat and storage power applications.

There are numerous systems for solar thermal heat storage that have been demonstrated more recently. They include two tank direct, two tank indirect, single tank thermal storage systems, thermal storage media (such as concrete thermal storage tested by The German Aerospace Center (DLR)), and thermal phase change materials.

It is important to understand that most thermal storage systems for solar farms will only really be needed many years from now. Because of the day and nighttime fluctuation in electricity demand, most electric generating demand is during daylight hours. As the capacity of solar farms grows beyond what is needed to begin to satisfy peak daytime load demands, only then will thermal storage systems begin to be needed for solar farms.

Thermal heat storage systems will continue to improve to further bring down costs and expand capacity when they are needed in the future. Because of the low cost of thermal storage systems compared to batteries or other electricity storage technologies such as fly wheels, it is expected that thermal storage will be the primary solar electric power storage choice for the foreseeable future.

A good example of the economics of storing solar thermal energy compared to battery storage is shown by comparing the energy stored in a laptop battery and a thermos. The battery in a laptop computer and a thermos bottle store about the same amount of total energy. The laptop battery costs about $150 and the thermos bottle costs about $4 wholesale.

We now see how base load solar farms with thermal storage and backup natural gas, PV's, wind generated electricity, hydro-electric and legacy operating coal and nuclear power plants are capable of providing 100% of the electricity and much of the transportation fuel needed to power America. As you will see in the next chapter, solar and renewable power can even be transmitted cost-effectively to regions such as New England that are a great distant from the arid desert regions where solar thermal farms will be located.

Other storage options such as hydrogen electrolysis will no doubt eventually emerge and these technologies can be integrated with thermal storage to produce electricity 24/7. As solar farms grow their share of base load power, the public may demand decommissioning of older nuclear and coal power plants. Demands for denial of future permit applications for coal and nuclear power will also likely grow louder.

EV's and their batteries are another source of electrical power storage supply. In the event of power outages, EV owners will have the ability to tap their EV batteries to supply their household electrical needs for many hours and even overnight. Utilities will also be able to contract with consumers and businesses to tap available EV batteries connected to the grid for peak load leveling, much as is done now by agreements between utilities and businesses to reduce usage during peak load alerts.

Many utility agreement programs allow for reducing the cost of electricity to businesses throughout the year in exchange for businesses agreeing to reduce usage within 20-30 minutes when the utility issues a peak load alert. Public utilities like Pacific Gas and Electric have also announced their intention to offer smart utility meters that will automatically select the time of day that has the lowest electric rate to charge an EV's battery. These smart meters will allow the utilities to more efficiently use their grids to level-out the amount of power that is supplied to their consumers, thereby reducing the overall delivery cost.

The EV load leveling capacity of a city or regional utility using these cost reduction agreements is enormous. The efficiency of long distance HVDC lines will even allow load sharing among different geographic regions. Integrating the power storage capacity of EV's with the electric grid allows utilities to operate more efficiently in ways that are currently not possible with liquid fuel based transportation systems.

Stored electrical power from an EV battery for portable applications is a further benefit of the EV revolution. At a campsite, for example, rather then burning white gas or propane for camp lighting, an EV can easily provide lighting for a family at night for 4 hours without significantly draining the EV's battery.

Advances in hydrogen gas generation from electricity will also provide a potentially large source of energy storage from large solar thermal farms and wind farms. As the cost of hydrogen gas generation decreases for water electrolysis using electricity, there will be much greater opportunities to store solar electricity generated power as hydrogen gas fuel for fuel cell EV's, backup electrical generation fuel, or for industrial process applications. The hydrogen gas can be generated either near the solar farms and distributed via pipelines, or generated directly at storage facilities, homes, or fueling stations. Hydrogen storage depots can be located around the country and serve to provide backup power and fuel sources where ever needed. The distributed nature of the hydrogen generation facilities would also serve to decrease the impact of short term power disruptions in the case of the temporary failures from the HVDC or HVAC grids.

Another example of the potential for decreasing the future cost of hydrogen gas generation through water electrolysis is the discovery by Dr. Daniel Nocera, head of M.I.T.'s Solar Revolution Project. Dr. Nocera and postdoctoral fellow Matthew Kanan discovered a method for reducing the cost of hydrogen gas generation from electricity by adding the metals cobalt and phosphate to water as part of a hydrogen electrolysis research program. Cobalt and phosphate cost about $2.25 and $.05 an ounce, in comparison to the currently most commonly used catalyst, platinum, which costs nearly 1000 times as much.

The potential of solar farms and PV systems to generate hydrogen gas will add another long distance electricity power storage option and another plentiful zero-emission transportation fuel source to a future solar electricity base load infrastructure.

7

Transcontinental Solar Electricity and HVDC

The movement of transportation fuel over the electric grid will have many benefits beyond eliminating vast numbers of fuel trucks from highways and oil tankers from our oceans and waterways. The electric grid will assume a new role as the transportation refueling superhighway. As a new HVDC grid is built and electricity capacity rises from solar and wind farms, the grid will have the capacity to transmit power across the U.S., Canada and Mexico.

In Europe, solar thermal heat storage generated electricity is already being produced in southern Spain. Consider the infrastructure that solar farms, a HVDC grid and EV's can replace over time. Significant reductions in the number of oil refineries, pipelines, tanker trucks, storage depots, and oil tanker ships from the Middle East and elsewhere can take place. We can eliminate military bases and Navy deployments required to protect oil shipping lanes, foreign production facilities, and vast coal strip mines and mile after mile of railroad coal boxcars.

One characteristic of a new HVDC grid will be matching regional and national periods of highest demand with the highest availability of solar electricity. Currently the highest load demands on the electric grid occurs in the daytime afternoon and in summer the peak is well pronounced in the afternoon hours to satisfy air conditioning demand. These peaks also correspond within a few hours of the peak capacity of solar thermal farms and PV electricity.

When the United States was originally electrified, the battle between DC (Direct Current) and AC (Alternating Current) was decided in favor of AC. This is why high power lines today in the U.S. are primarily AC. A downside to AC high power transmission lines is the relatively high electrical losses over long distances due to electrical resistance in the lines and other effects. The potential for longer range and more efficient HVDC power transmission is not well known in the United States.

The modern AC transmission system in use today was developed in the late nineteenth century by Nikola Tesla (1856 – 1943), an inventor and physicist. He was born in Croatia which at that time was part of the Austrian Empire. He later studied electrical engineering at the Austrian Polytechnic in Graz (1875), and eventually moved to the United States and became an American citizen.

Tesla was hired by Thomas Edison to work at the Edison Machine Works in New Jersey. While employed by Edison, Tesla solved many of Edison's most difficult problems and filed numerous patents while under the employ of Edison. Tesla eventually resigned and formed his own company, Tesla Electric Light & Manufacturing.

He eventually went on to create and patent a system of generating alternating current (AC) over long distances. He also invented induction motors which ran on AC and ultimately made AC more practical for building a commercial power grid infrastructure.

Tesla ultimately sold his patents for his "polyphase" AC generation technology to George Westinghouse, who began to promote AC power through Westinghouse's company.

AC systems overcame the limitations of the direct current system used by Thomas Edison to distribute electricity efficiently over long distances. The turning point in the battle between DC and AC high power transmission in the U.S. came to a head in 1893 with the opening of the Chicago World's Fair. This competition between the DC of Thomas Edison and the AC of Nikola Tesla and George Westinghouse was known as the War of Currents.

Westinghouse Corporation and Tesla's AC system won the competition for lighting the Chicago World's Fair, the world's first all-electric fair. The fair was also known as the Columbian Exposition —celebrating the 400th anniversary of Columbus discovering America. Edison's company, General Electric, and Westinghouse bid on a contract to light the Columbian Exposition. Westinghouse's million-dollar bid was lower then General Electric's by nearly 50%. Much of the reason for GE's higher bid was due to the amount copper wire needed for DC power distribution using equipment available at the time of the 1893 World's Fair.

The Columbian Exposition opened on May 1, 1893. President Grover Cleveland traveled to the Chicago World's Fair and was given the honor of turning on the power for the Exhibition. The excitement built as President Cleveland pushed a button and one hundred thousand lights brilliantly lit the fairground's buildings with a blaze of white light. The Columbia Exposition became known thereafter as the "City of Light". In the Fair's Great Hall of Electricity, the Tesla AC generator was seen by millions of visitors. From that time forward, it was clear that Tesla's AC system would become the backbone of the electric grid for America into the 20th century.

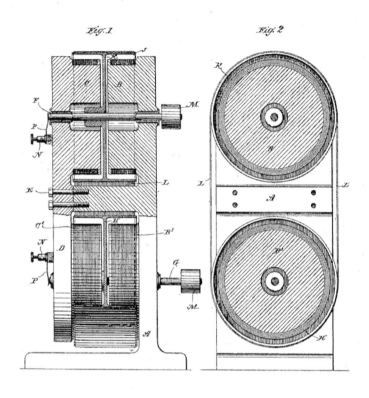

(No Model.)

N. TESLA.
DYNAMO ELECTRIC MACHINE.

No. 406,968.   Patented July 16, 1889.

Fig. 1   *Fig. 2*

Witnesses:
Robt. F. Gaylord
Ernest H. Hopkinson

Inventor
Nikola Tesla
by
Duncan, Curtis & Page.
Attorneys.

M. PETERS. Photo-Lithographer, Washington, D. C.

Nikola Tesla's AC dynamo used to generate AC which is used to transport electricity across great distances.

The first modern commercial power plant using the new alternating current generators was at the Mill Creek hydroelectric plant near Redlands, California in 1893 designed by Almirian Decker. Decker's design incorporated 10,000 volt three-phase transmission and set forth the standards for a complete system of generation, transmission and motors still in use today.

It was not until decades after AC became a well established transmission standard for the U.S. that a practical technology for HVDC became available using mercury arc valves. A mercury arc valve (mercury vapor rectifier) is a type of electrical rectifier that converts alternating current into direct current. Mercury arc valves were invented by Peter Cooper Hewitt in 1902 and further developed throughout the 1920s and 1930s by researchers in both Europe and North America. Mercury arc valves first enabled practical HVDC power transmission to begin to be used commercially in the 1920's.

HVDC transmission technology continued to be developed more extensively in the 1930s in Sweden at the ASEA Corporation. Early commercial installation examples of HVDC included a 30 MW (Megawatt) line in the Soviet Union in 1951 between Moscow and Kashira, and a 10-20 MW system in Gotland, Sweden in 1954.

Fully-static mercury arc valves began to be used commercially for HVDC transmission in 1954. Mercury arc valves were common in systems designed up until the 1970's. HVDC systems after the 1970's used more advanced solid-state power rectifier devices.

The advantage of high voltage direct current in more modern times is its ability to transmit large amounts of power from remote generation sites over great distances with high transmission efficiency and to interconnect non-synchronized alternating current power grids.

High voltage is used in long distance power transmission to reduce the energy lost due to resistance of the transmission lines. For a given quantity of power transmitted, higher voltage reduces the transmission power loss because power in a circuit or power line is proportional to the current, but the power lost as heat in the wires is proportional to the square of the current. Power is also proportional to voltage, so for a given power level, higher voltage can be traded off for lower current- the higher the voltage, the lower the power loss.

Power loss can also be reduced by reducing the line resistance, commonly achieved by increasing the diameter of the conductor line; but larger conductors are heavier and more expensive. High voltages cannot be easily used in lighting and motors, so transmission-level voltage must be reduced to values compatible with end-use equipment. A transformer, which only works with alternating current, is the most efficient way to change voltages.

Practical manipulation of DC voltages only became possible with the development of high power electronic devices such as mercury arc valves and later semiconductor devices including thyristors, IGBTs, MOSFETs and GTOs.

HVDC is well suited for interconnection of two different AC regions of a power-distribution grid, as would be needed for long range solar farm power transmission.

The conversion from AC to DC is known as rectification, and from DC to AC as inversion. Beyond about 50 km for submarine cables and 600–800 km for overhead cables, the lower cost of the HVDC electrical conductors begins to outweigh the cost of the conversion electronics required between HVDC and HVAC.

Additional advantages in the use of HVDC links between HVAC networks are the increased stability and capacity in a combined transmission grid.

There are now over 100 HVDC lines in operation worldwide. Depending on the voltage level and construction details, losses are generally about 3% per 1000 km for HVDC plus the losses from conversion between DC and AC.

The Itaipu HVDC Transmission Project in Brazil, owned by Furnas Centrais Elétricas S.A. in Rio de Janeiro (an Elétrobras company), is currently the most impressive HVDC transmission line in the world. It has a total rated power of 6300 MW and a voltage of ±600 kV DC. This is the equivalent electrical power of 6 large nuclear power plants. The Itaipu HVDC transmission system consists of two bipolar DC transmission lines bringing power generated at 50 Hz in the 12600 MW Itaipu hydropower plant to the 60 Hz network in São Paulo, the industrial center of Brazil. Power transmission started in October 1984 with 300 kV and in July 1985 with 600 kV. The converter stations were commissioned sequentially in order to match the generating capacity build-up at the Itaipu hydropower plant.

HVDC systems also make possible the interconnection of unsynchronized AC networks. Many of the HVDC transmission links currently in worldwide use are primarily for transmission of power from remote generation sites. Wind farms located off-shore are also considered prime customers for use of HVDC systems to collect power from multiple unsynchronized generators for transmission to the shore by underwater cable.

An optimized long distance HVDC transmission line has lower losses than AC lines for the same power capacity. There are also losses in the converter stations, which are about 0.6 % of the transmitted power at each station. The total HVDC transmission losses are therefore lower than the AC losses in longer distance transmission applications. HVDC undersea cables also have lower losses than HVAC undersea cables. The diagram below shows a comparison of the losses for overhead line transmissions of 1200 MW with HVAC and HVDC transmission lines.

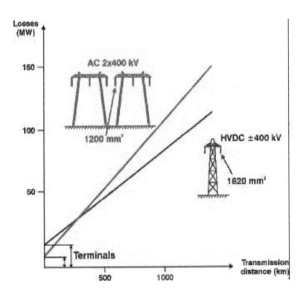

Transmission Losses including 2 way conversion (ABB)

HVDC lines are therefore capable of supplying electricity from southwestern solar thermal farms (with overnight thermal storage) to a New England home on a cold December night with only moderate electrical losses. Solar power generated 24 hours a day can therefore be distributed to all major parts of the continental U.S. reliably using a HVDC backbone with connections to the existing HVAC grid. The same will be true for other regions in the world including North Africa to Europe, Australia and Western China's deserts to the populated eastern Chinese cities.

The transition to this new grid infrastructure will be incremental and very complementary to the existing grid infrastructure. As more solar farms and PV installations occur, increasing amounts of electrical capacity can be supplied. This will reduce and then reverse the need to build additional coal, natural gas, or nuclear power plants. Windmills and wind farms will add capacity where it is cost effective, reliable and permitted by regulatory agencies. As the capacity of solar farms grows beyond what is required in the surrounding regions, the backbone for a new longer distance HVDC grid will emerge given regulatory support. Over time, as more and more of the HVDC grid is built, it can begin to gain the capacity to carry electricity nationally from southwestern solar farms, distributed rooftop and smaller solar farms and regional wind farms.

HVDC power lines can also have decreased right-of-way requirements for land use, often requiring only a single line of towers instead of a double set in the case of comparable HVAC transmission systems.

At the terminus of the HVDC transmission lines, the electricity is then converted back to AC and distributed through the existing AC transmission systems to the end users.

Other uses of direct current today includes charging batteries and as the power supply for many electronic systems. Very large quantities of direct-current power are also used in production of aluminum and other electrochemical processes.

Expansion of the long distance HVDC grid backbone will include multiple redundant paths due to the need for high reliability. Downed power lines, fires, earthquakes, terrorism, and mechanical failure are all factors in determining the degree of redundancy needed. The increased usage of the electric grid for transportation power will facilitate investments in grid redundancy.

Compared to the cost of oil refineries, pipelines, tanker trucks, service stations and supertankers needed to support the current transportation fuel infrastructure, the cost for a robust long distance HVDC grid linking most major areas of the United States to solar and wind farms nationwide would be relatively modest.

How much of the U.S. electric generating capacity can be supported by solar farms? Coal currently supplies about 53% of U.S. electrical power generation. It would require several decades for solar power to achieve this level of capacity in the U.S., but the arid land resource base and rooftop space is available. By that time, the reliability of solar farms will be proven, and the technologies will have improved to where it will be possible to generate in excess of 50% of the electric power in the U.S. from solar thermal farms with overnight storage, natural gas and hydrogen backup, and photovoltaics. New electric capacity additions will also come from wind, and other renewables.

Combined with geothermal and existing hydro-electric power, and the declining production from decommissioning coal and nuclear power plants, we now can see a mix of power resources that can satisfy energy demand through the end of this century with major reductions in greenhouse gas emissions, fossil fuel dependency and air pollution.

As natural gas power plants are idled, many of them can be retained as backup power generators for use during times when solar farms are not able to meet overall demand. Many natural gas power plants can be turned on and off for power generation, unlike coal power plants that are not well suited for this role. Many natural gas power plants are used today in this "peak power" role frequently for daytime and peak summer air conditioning demand.

The growth of electrical charging stations can be very rapid with the conversion to electric transportation. Charging stations will be installed in many more locations then gas stations and require very little maintenance. Also, charging stations will have none of the damaging environmental side effects of contaminated ground water from leaking gasoline service station storage tanks that have plagued many communities.

There are 2 primary classes of re-charge stations. High power stations for the high re-charge rate fast charge batteries for EV's and low power re-charge stations for slower charging and smaller vehicles including electric bicycles.

Fast charging high power stations are capable of re-charging fast charge technology EV batteries like the A123

Systems batteries very quickly using charging systems that have been sold commercially for many years.

Lower power recharge stations are less expensive and can be more widely deployed throughout the country and can be as simple as your home electrical 110 volt or 220 volt power outlet. Payment at charging stations can be similar to public telephones that allow ID numbers and pin codes, credit card swipes, or even wireless cell phone proximity payment that is common in Japan for consumer retail purchases. Charging stations can also be built-in at gas stations for the convenience of plug-in hybrid refueling of both gas and electricity for travelers.

A national HVDC grid will also allow backup power plants to be brought online from distant locations and supply power nationwide during times of excess cloud cover in the base load solar farm regions. One of the reasons solar farms are capable of carrying base load demand is their very high availability and scale to supply the grid and low cost overnight storage. Wind power is much less reliable and cannot supply nearly the same level of reliable power as solar farms in the U.S. Southwest, North Africa, Australia, and many and the arid parts of Asia.

What happens when a motorist runs an EV battery down completely and is stranded on the road side? There are numerous simple solutions for wayward EV drivers who forget to "power up" before or during a trip. In the case of a plug-in hybrid, this is not a problem since the car can seamlessly tap the gas or diesel engine and supply power to the car. Tow trucks can be outfitted with chargers and small portable generators can be made available to borrow at gas stations and other roadside locations. There is no reason EV's need to be anymore at risk of running out of fuel then gas or diesel powered engines.

Investments in a highly redundant HVDC long distance grid backbone combined with upgrades to existing HVAC systems will allow the U.S. to provide clean, renewable electrical and transportation power nationally and replace tens of thousands of miles of coal boxcars, billions of barrels of oil, and trillions of cubic feet of natural gas every year.

Adding electrical power fed back from EV's plugged into the electrical grid is also an option to increase the reliability of the power grid with very large scale backup power. Pacific Gas and Electric has already demonstrated the ability of electric vehicles to feed power back to the grid with Vehicle-to-Grid (V2G) technology. V2G technology allows for the bi-directional sharing of electricity between EV's, Plug-in Electric Hybrid Vehicles and the electric grid. The technology turns each vehicle into a power storage system, increasing power reliability and the amount of renewable energy available to the grid during peak power usage.

The technology is flexible enough to allow EV owners to select a price threshold at which they are willing to sell energy, and when the price reaches this point the utility would be able to automatically draw energy out of each participating vehicle, leaving enough for the drive home if necessary. The utility's customers would earn credit for the amount of energy used by the utility toward their monthly energy bill.

8

Solar Water Heating Rebirth

Few people in the United States realize that solar water heating for residential use has gone through several cycles of popularity and descent into obscurity. Solar water heaters for hot water are one of the simplest solar technologies and one of the oldest.

As natural gas, electricity and heating oil prices rise, once again solar water heating in America will receive more attention. How practical solar water heating is for any residence depends largely on where you live, the building you occupy, the local sun exposure and the annual regional amount of sunshine.

For people living in southern or sunny climates, solar water heaters can provide economical hot water and freedom from future increases in water heating bills. You also will be adding to your list of non-fossil fuel dependencies and lowering your carbon footprint.

Israel and Spain are leaders in the use of solar water heaters. Following the 1970's energy crisis, the Israeli Knesset required all new homes to install solar water heaters. In 2006

Spain became the second country in the world to require solar water heaters for new construction. Solar water heating on Spanish buildings must now provide 30-70% of a building's water heating needs, depending on the climate zone, consumption level and back-up fuel availability.

Residential solar water heating also became popular from the late 1890's through the 1920's in Southern California, Arizona, New Mexico and Florida. During this time the technology of solar water heaters improved dramatically. Solar water heaters developed in the 1890's were not able to maintain hot water overnight and suffered from significant heat losses.

By the 1920's, solar water heaters had become much more refined and were capable of providing hot water 24 hours a day during sunny weather, and could be integrated with external heating sources and provided guaranteed hot water year round. In the 1920's the boom in cheap natural gas production began along with free natural gas water heater installations. This combined with easy credit terms put an end to the solar water heater industry in the southwestern U.S. by the 1930's.

The technology behind solar water heaters today has come a long way since the early 1900's. Sunny climates offer a home owner a cost-effective way to supplement fossil fuel or electricity based hot water heating. You also have the opportunity to further reduce your dependence on future fossil fuel price increases, and once again move away from the need to conserve energy because of cost or environmental concerns. You will be able to wash clothes and take showers and baths knowing that most of your future cost of hot water is free after you have paid for your solar water heater.

A typical residential solar water-heating system reduces the need for conventional water heating by about two-thirds in favorable climates.

Active Indirect System

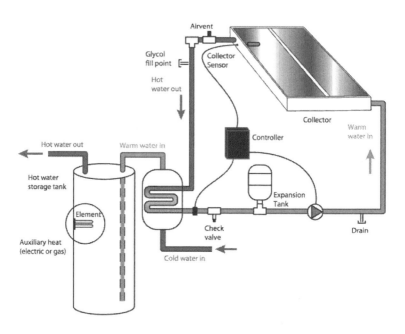

A Typical Rooftop Active Indirect Solar Water Heating System

Warm weather locations, where freezing protection is not a problem, allow solar hot water heaters to be very cost effective. There are additional design requirements for colder weather, which can add to the cost of a solar water heating system. This has the effect of increasing the initial cost of a solar hot water system to a level higher than comparable conventionally produced hot water.

When calculating the total cost to own and operate a solar water heater, the analysis will take into consideration that solar energy is free, thus greatly reducing the operating costs, whereas other energy sources, such as gas and electricity, can be quite expensive over time. Thus, when the initial costs of a solar system are properly financed and compared with the total energy costs, in many cases the total monthly cost of solar heat can be less than the total life-cycle cost of conventional natural gas or electric hot water heaters. In addition, federal, state and local incentives can significantly reduce the total overall cost of solar water heaters.

As an example, a 56 square foot solar water heater can cost $7,500, but that initial cost can be reduced to just $3,300 in the state of Oregon due to federal and state incentives. The system will save approximately $230 per year, with a payback of 14 years or less, depending upon how fast local natural gas or electricity rates rise over time.

Lower payback periods are also possible if solar water heaters are part of new home construction. Southern or arid climates allow faster payback periods. However, in more northerly locations, solar heating is less efficient. Useable amounts of domestic hot water will be available mostly in the summer months, or on cloudless days between April and October. During the winter and on cloudy days the output can be very limited in northerly climates.

Solar water heaters use the sun to heat either water or a heat-transfer fluid in the collector. Heated water circulates and is then held in the storage tank and is ready for use, with a conventional system providing additional water heating as necessary. The tank can be a modified standard water heater tank, but it is usually larger and very well insulated. Solar water heating systems can be either active or passive, but the most common are active systems.

Active Solar Water Heaters

Active solar water heaters rely on electric pumps and controllers to circulate water or other heat-transfer fluids through the collectors. Below are the three types of active solar water-heating systems:

1. Direct-circulation systems using pumps to circulate pressurized potable water directly through the collectors. These systems are appropriate in areas that do not freeze for long periods and do not have hard or acidic water.
2. Indirect-circulation systems pump heat-transfer fluids through collectors. Heat exchangers transfer the heat from the fluid to the potable water. Some indirect systems have "overheat protection," which is a means to protect the collector and the glycol fluid from becoming super-heated when the load is low and the intensity of incoming solar radiation is high. The two most common indirect systems are:

a. Antifreeze. The heat transfer fluid is usually a glycol-water mixture with the glycol concentration dependent upon the expected minimum temperature. The glycol is usually food-grade propylene glycol in order to be non-toxic.

b. Drainback systems. A type of indirect system which uses pumps to circulate water through the collectors. The water in the collector loop drains into a reservoir tank when the pump stops. This makes drainback systems a good choice in colder climates. Drainback systems must be carefully installed to assure that the piping always slopes downward, so that the water will completely drain from the piping.

Passive solar water heaters

Passive solar water heaters rely on gravity and the tendency for water to naturally circulate as it is heated. Because they contain no electrical components, passive systems are generally more reliable, easier to maintain, and possibly have a longer working life than active systems. The two most popular types of passive systems are:

1. Integral-collector storage systems consisting of one or more storage tanks placed in an insulated box with a glazed side facing the sun. These solar collectors are suited for areas where temperatures rarely go below freezing. They are also good in households with significant daytime and evening hot-water needs; but they do not work well in households with predominantly morning draws because they lose most of the collected energy overnight.

2. Thermosyphon systems are an economical and reliable choice, especially in new homes. These systems rely on the natural convection of warm water rising to circulate water through the collectors and to the tank (located above the collector). As water in the solar collector heats, it becomes lighter and rises naturally into the tank above. Meanwhile, the cooler water flows down the pipes to the bottom of the collector, enhancing the circulation. Some manufacturers place the storage tank in the house's attic, concealing it

from view. Indirect thermosyphon systems (that use a glycol fluid in the collector loop) can be installed in freeze-prone climates if the piping in the unconditioned space is adequately protected.

A Traditional Solar Water Heating System

Solar Water Heaters for Pools

Using the sun to heat a swimming pool is a very effective way to use solar energy. Solar pool heaters can be connected to the pool's existing water circulation system. They cost anywhere from $2,000 to $5,000, need very little maintenance and can allow people to swim during months that are otherwise too cold to swim.

Collectors: Sizing and Orientation

A typical solar pool heater consists of a collector that is made of plastic panels. The panels have tubes (called headers) on the top and bottom of the panel that allow water to flow into and out of the plastic panel. The headers are connected by many small tubes through which water flows and gets heated by the sun.

Solar pool covers can also be combined with pool solar water heaters. Solar pool covers cost about 50 cents per square foot and can last up to five years, although I usually replace mine about every 2-3 years. A translucent cover works better than a dark or opaque cover because it allows sunshine to warm the pool through the plastic covering. Placing a cover over the pool can raise water temperatures between 5F and 15F.

The actual amount of temperature increase depends upon the amount of sunny weather, the outside temperature and the humidity. One time I went on vacation with my solar heating system turned on and the pool cover left in place. When I returned 2 weeks later, the entire pool was 99 degrees F. The size of a collector needed for a swimming pool depends on several factors including the size of the pool, the climate, the sun exposure of where the solar pool water-heaters are placed and the desired water temperature.

Solar Panels

Normally, the total area (square footage) of the solar collector should be at least half of the pool surface area. For example, if the pool covers 500 square feet, the collectors should be at least 250 square feet.

Collectors should face south and be tilted at an angle equal to the latitude of the pool's location minus 10 to 15 degrees. If this is not possible and the collectors must be laid flat or must face west, the collector will not get as much sunlight. In this case, a larger collector area may be needed to make up for the decrease in collector efficiency resulting from less sunlight.

Pool collectors can either be mounted on the roof of a building or mounted on a frame on the ground near the pool. Where the collectors get placed depends on how much space is available and how much sunlight shines on the space.

A Typical Pool Solar Panel Configuration

Pool Covers

Pools lose heat through convection, evaporation, radiation from the pool surface, and by conduction heat loss from the sides or walls of the pool. The amount of heat that is lost depends on many factors, such as the water temperature, air temperature, humidity, and winds blowing across the pool surface.

On average, outdoor pools lose almost 90% of their heat from the water surface: 70% by convection and evaporation, which are related, and 20% by radiation to the sky. About 10% is lost by conduction from the sides and bottom. The best way to prevent the heat loss from evaporation is with a plastic pool cover. How much a pool cover in sunlight can heat a pool depends on the type of cover, when and how long it is used during the day, the season and local climate and how many hours per day of sunlight the pool gets. At my house in California, the south facing rooftop solar water heater and pool cover allowed my pool to maintain a temperature between 85 degrees and 96 degrees from April to October each year, and the total cost of heating the pool is $0.00, excluding the modest cost of electricity required for the pool pump.

9

Nuclear Power- A Long Term Policy Mistake

What's not to like about nuclear power plants? Besides the fact that they produce deadly waste, the power plants are deemed by the National Academy of Sciences as terrorist targets, and the waste must be cooled for 40 years. Then the highly radioactive waste must be shipped across the country, and buried in extremely expensive storage depots for up to a million years. Nuclear power plants are also radioactive after their 40 to 60 years of operation and have to be disassembled and buried under expensive radioactive protocols.

Nuclear power plants have always been controversial, with good reason. In this chapter we'll examine what the risks and benefits are in comparison to solar power for base load electricity generation. There are many unique risks with nuclear power.

They include the risk of a power plant catastrophic accident, terrorist attack, secure storage of spent nuclear fuel in cooling ponds for decades, long term nuclear waste storage (for up to a million years), transporting nuclear waste across the country, and competition for uranium supplies.

The cost alone for the only U.S. high level nuclear storage site, Yucca mountain, for 100 years was estimated in 2000 to be $58 billion (year 2000 dollars), not including currently escalating cost overruns. Since 2000 estimates have risen to over $90 billion.

Nuclear power originally was offered as a large scale, safe, inexpensive long term electricity supply source far into the future. The reality proved to be somewhat different. The nuclear power industry was looking to commercialize the technology of nuclear power originally developed for nuclear weapons. The public was told that the industry could virtually guarantee that a nuclear accident was all but impossible and that fuel disposal would not be a problem.

One example of the faulty assumptions at work was the Three Mile Island nuclear power plant accident. Three Mile Island was put into service on September 2, 1974 in Harrisburg, Pennsylvania. On March 28, 1979, the power plant "malfunctioned", releasing radiation into the surrounding environment. The accident almost resulted in a widespread catastrophic release of radiation that would have contaminated large areas around Harrisburg for generations and resulted in tens of billions of dollars in cleanup costs and hundreds of billions of dollars in economic losses.

The nuclear accident at Chernobyl in 1986 in the former Soviet Union country of Ukraine shows the risks of a full scale nuclear accident. Vast areas of land surrounding the power plant are off limits to human habitation for many generations, billions of dollars in losses and the risk of further contamination for *thousands* of years. The main argument in favor of nuclear power plants after Three Mile Island and Chernobyl is that we have learned from our mistakes and it will not happen again.

The fundamental flaw with this type of argument is illustrated in the last 2 NASA Space Shuttle accidents. The first accident occurred in 1986 when the Space Shuttle Challenger exploded 73 seconds after launch. The disaster was found to be the result of a faulty design in the O-rings of the solid rocket boosters that were attached to the Challenger. After extensive analysis, studies, panel hearings and redesign, the Shuttle Program launches were re-started with much improved safety standards in place.

On February 1, 2003 the Space Shuttle Columbia disintegrated over Texas upon re-entry due to damage sustained to the external heat shield tiles during lift-off. Once again the Space Shuttle program suffered a catastrophic loss due to designs and circumstances that had not been anticipated since the first accident. The take away lesson is that fixing one failure mode with aging hardware doesn't guarantee future disasters won't occur. Fortunately NASA's mistakes will not result in large sections of the U.S. becoming uninhabitable for centuries.

It is sadly ironic that part of the resurgent interest in building more nuclear power plants is the fear of global warming and increased CO_2 emissions. How significant the threat of global warming is from increasing CO_2 emissions is actually not well understood scientifically. This scientific uncertainty has largely been lost in the debate regarding the potential policy solutions.

Since CO_2 reductions from new power plants are deemed to be in mankind's best interest, virtually any energy source that does not emit CO_2 is, by some policy maker's criteria, deemed to be beneficial. Why would professed environmental organizations and policy makers encourage more nuclear power, whose highly radioactive waste is a threat for hundreds of thousands of years, when even the worst possible effects from increased CO_2 emissions would last for less then a few hundred years, and potentially not exist at all as an environmental threat for the next 50 years or more?

A major part of the "green" push for nuclear power comes from the nuclear industry lobby. The combined impact of lobbyists, and the argument that nuclear power is a "green" solution has opened up the possibility in the U.S. to significant numbers of new nuclear power plant permits being approved in the next several years. This trend combined with federal incentives and tax credits is encouraging utilities under the threat of carbon emission caps to propose building new nuclear power plants.

Recent cost estimates for nuclear power plants have more then doubled and are still rising. The Florida utility, FPL Group, recently estimated that the cost of building a new nuclear power plant in southeastern Florida will be up to $9 billion dollars, more then double previous estimates. Many states are clearing a path for nuclear power plant development before the full costs are even known due to global warming politics.

New nuclear power plant proposals can take 7-10 years from submission to completion. Compare that to an approximately 3 year timeframe for permitting and constructing a large solar thermal farm.

So now we have another set of potential disasters that may arise from the politically popular knee-jerk reaction to climate change fears by policy makers who receive nuclear lobbyist money. We have witnessed food prices increase and risks of global famine due to biofuel mandates literally burning up part of our food supply. We are now also seeing the rapid movement towards increasing use of nuclear power which will inevitably lead to more nuclear waste production and the risk of catastrophic accidents or terrorism that may release radiation that can persist for many centuries into the future.

"The National Academy of Sciences has identified that the nation's current nuclear reactor waste storage system is vulnerable to deliberate attack," said Paul Gunter of the Nuclear Information and Resource Service and spokesperson for the citizens' coalition. "Robust on-site storage of nuclear waste — hardened against rocket propelled munitions or explosive laden aircraft — provides the public with the first responsible steps towards a protective strategy from a nuclear waste fire that could induce tens of thousands of cancer fatalities out hundreds of miles," said Gunter.

The risk of a nuclear accident or terrorism in the United States will only rise with time. As nuclear power plants age, metal fatigue in and around the nuclear core begins to elevate the risk of a materials failure. Many nuclear reactors in the U.S. are approaching their 40 year operating license lifetimes. The nuclear power industry is submitting applications to extend the permitted lifetime of these power plants as well as applying for new nuclear power plant permits.

It is to be expected that the nuclear industry will attempt to continue to operate nuclear power plants well beyond their designed lifetimes. It is also clear that as the nuclear power plants age, the risks and unknowns associated with aging nuclear cores and materials failures will elevate the risks of accidents that were not anticipated. In all likelihood existing old nuclear power plants may be approved for re-licensing until there is another nuclear accident due to an aging nuclear reactor from a failure mode that wasn't anticipated. Of course if there is another nuclear power plant accident, it will be heralded as an accident that "could not have been anticipated".

According to a report of experts (Lamb & Resnikoff, 2001), a severe rail incident involving nuclear transported waste, with accident conditions similar to the Baltimore rail tunnel fire in July, 2001 could cause thousands of cancer deaths, and cost $10-$14 billion in clean-up costs. According to a 1985 DOE study, a similar accident in a rural area would contaminate 42 square miles (an area roughly the size of Washington, DC).

One of the biggest problems with policy discussions regarding the true cost of nuclear power is accurately estimating the cost of decommissioning nuclear power plants. In Britain the cost of cleaning up ageing nuclear facilities, including some described as "dangerous", has risen above £73bn ($145 billion) according to Britain's National Audit Office. Nineteen sites across the country, some dating from the 1950s, are due to be dismantled in the coming decades.

Compare the cost of decommissioning a nuclear power plant and storing the nuclear waste for more then 100,000 years to the cost of simply maintaining and refinishing the reflective surfaces of solar thermal farms. Add to the cost of providing security for the nuclear power plants, the cost of waste storage, disaster insurance, waste transport infrastructure and long term storage depots.

The decision needs to be made <u>now</u> not to license new nuclear power plants in the United States. Licensing more nuclear power plants will lead to a larger nuclear waste infrastructure that inevitably will likely lead to future nuclear accidents, leaks or worse and a much less safe world for millennia. The special interest subsidies, incentives and insurance guarantees that benefit the nuclear industry should also to be ended.

In July 2008, another of the inevitable accidents occurred in southern France. France's nuclear safety watchdog ordered a nuclear power plant in Provence to temporarily close after a uranium leak polluted the local water supply. Waste containing uranium leaked into two rivers at the Tricastin plant at Bollene,

40km (25 miles) from the popular tourist city of Avignon. People in nearby towns were warned not to drink any water or eat fish from the rivers since the leak occurred. Officials also cautioned people not to swim in the rivers or use their water to irrigate crops.

There has been much discussion recently concerning the re-processing of spent nuclear material. Although re-processing can reduce the amount of nuclear waste that must be contained in long term storage, it cannot eliminate nuclear waste. It also creates a larger infrastructure of nuclear fuel, processing, transportation and waste. Ultimately, the risks of accidental or deliberate high level nuclear contamination or catastrophic accidents will only rise over time. It is unfortunate these risks are even being considered in the face of safe, economic solar electricity that can fill our future power needs through this century.

Nuclear power should be understood for what it really represents- a highly risky long term threat to a safe environment that risks centuries of unnecessary radioactive contamination over large inhabited land areas.

10

The U.S. Biofuels Money Pit

What role should U.S. produced biofuels play in the short and long term as a significant transportation fuel source for the U.S.? Governments around the world have already embarked on policy changes that are making biofuels a significant source of transportation fuel because of higher oil prices, policy mandates and subsidies.

There are a number of different biofuels, biofuel sources and production methods. In this chapter we'll examine why U.S. grown food crop based biofuels pose a serious risk to the global food supply and may be a source of large scale famine risk and why for the U.S., biofuels may do more harm then good.

The main sources for biofuels for the short, medium and longer time horizons for the U.S. will likely be corn and cellulosic ethanol feedstocks. Some Asian countries like Indonesia produce large and increasing amounts of palm oil to be used for biofuel. Much of the palm oil is used in Europe and elsewhere to produce biodiesel. Palm oil is only practical to grow commercially on plantations in the tropics.

A consequence of Indonesia's rapid growth of palm oil plantations is resulting in large scale deforestation that also releases enormous quantities of CO_2 into the atmosphere. Indonesia is now the world's third largest CO2 emitter, behind only the U.S. and China, due largely to massive forest burning to clear land for more palm oil plantations.

Cellulosic ethanol is the conversion of the fibrous plant cellulose parts to ethanol. Cellulosic ethanol offers the potential to use plant materials other then food crops to produce ethanol.

One of the first considerations when trying to understand if biofuels make sense economically and environmentally are the regional differences in their production potential. Brazil is the largest ethanol producer in the world, and a substantial percentage of Brazil's transportation fuels use ethanol. Brazil's ethanol production is based on sugar cane, which is 7 times more efficient then corn in extracting useful energy for transportation fuel. In most businesses, a 700% difference in production costs usually means the difference between corporate success and bankruptcy.

Just because Brazil is able to grow and process sugar cane economically into ethanol in the country's tropical regions does not mean that corn ethanol makes any sense for the U.S. to grow as an economical biofuel. The specifics of corn ethanol's practicality in the U.S. must be assessed independently of Brazil's sugar based biofuels program.

The use of ethanol as a fuel goes back to the early 19th century. An engine that ran on turpentine and ethanol was developed in 1826.   Since then ethanol's popularity as transportation fuel has ebbed and flowed. It wasn't until 1974 that the U.S. Congress passed legislation to promote the use of ethanol for transportation fuel.

One of the biggest mistakes we can make is to assume that because Brazil's sugar based ethanol is economically viable and has reduced Brazil's dependence on fossil fuels, the same success will be derived from corn based ethanol in the U.S. and elsewhere.

Let's look at our 5 basic attributes of energy supply suitability for corn based ethanol and then understand why corn ethanol has such significant policy and subsidy incentives.

Is corn based ethanol a reliable energy source? Surprisingly the answer is no. One of the first impacts from the large subsidies and mandates for the use of corn ethanol in gasoline was the drawing down of the year-to-year corn and grain stockpiles that served to cushion yearly variations in abundant harvests and crop failures. The combination of high demand for corn as food and as an ethanol feedstock not only has raised the price of corn and other feed grains, but it also can draw down stockpiles to dangerously low levels.

Looking into the future, it is inevitable that some years will see either very poor corn harvests or worse, a region wide crop failure due to drought, floods, pests or other factors. Policy makers and industry will then be faced with a crop shortage for both fuel and foodstuffs. A national corn crop without large carry-over stockpiles is not reliable as a source of both food and fuel.

Having a large oil producer go offline for a few weeks is one thing, but having your country's transportation fuel source fail for at least a year is quite another. In fact it's blatantly poor energy policy from purely a reliability perspective.

It is also important to understand that corn ethanol only marginally produces more energy then is consumed to produce it when including the natural gas required for making nitrogen fertilizers, planting and harvesting fuel expenditures, shipping, production, refining, and transportation. It requires large amounts of fossil fuels to produce corn ethanol each season.

Is corn ethanol safe? It is generally considered safe. However, the massive use of nitrogen based fertilizers required for the expansion of corn farmland filters into the rivers and runoff and in the case of the Mississippi river, flows into the Gulf of Mexico and only makes the dead zone at the mouth of the Mississippi that much larger.

The net carbon and CO_2 saved is almost completely offset by the natural gas, transportation fuels and coal based electricity that is currently needed to run the ethanol refineries. Due to the characteristics of ethanol, it cannot be shipped via pipeline and therefore requires transportation vehicles and trains that burn fuel in order to transport the ethanol.

Is corn ethanol sufficiently scalable to become a primary fuel supply in the United States and elsewhere? Because corn ethanol is so energy inefficient, it requires large inputs of energy to make one gallon of ethanol. A 2005 ethanol energy input study concluded that corn requires 29 percent more fossil energy than the fuel produced. More recent studies claim that corn ethanol can be net energy positive, up to 25%, but the central problem is clear.

Corn ethanol produces little new energy, if any, beyond the energy required to make the ethanol. The result is that to produce the large quantities of ethanol required to impact substantially our transportation fuel needs, enormous amounts of land, fertilizers, water and power will be required. Each of these inputs is also a limited resource. Because of the large energy inputs required, the net carbon savings over gasoline or diesel is small for corn ethanol. Sugar based ethanol has much more favorable energy balance and carbon neutral advantages, but the U.S. does not have the climate advantages to grow sugar cane on the scale that Brazil has achieved.

The root of this seemingly misguided policy decision by the U.S. Congress and President to subsidize corn ethanol to the tune of $1.40 per gallon ($0.46 blender's credit plus other subsidies and supports) is largely the result of the "green" hysteria, special interest groups and a fundamental flaw in the U.S. Constitution.

The U.S. Senate consists of 2 Senators from each state. This means that Nebraska, a farm state that according to the USDA produces 10% of the U.S. corn crop, has the same voting power in the U.S. Senate as California. Nebraska has a population of approximately 1.7 million people and California has 37 million. In other words, Nebraska has 4.8% of the population as California, but has the same Senate voting power. This same distortion in voting leverage exists across most of the corn farming states. The selection of corn ethanol as a fuel substitute, with its poor net energy balance, large resource consumption, and poor economics was driven not by an efficient energy policy, but by a flaw in how the U.S. Constitution apportions power in the Senate. In essence the leverage for Nebraska Senators per state resident is 20 to 1 over that of California and other large primarily industrial or urban states.

This also explains how a federal policy could be drafted that in essence allows part of a country's food supply to be burned for fuel with little or no significant net benefits other then to increase the demand and price of corn to highly leveraged farm states.

Recently there have been calls for additional low income subsidies for food because of the high prices caused by "burning" grain crops for fuel. So now tax payers pay for subsidizing burning the food crops for fuel, with little net energy gain, and then pay for subsidizing the resulting high food prices. Bad policy begets more bad policy.

Other adverse impacts are the vast increase in ground water consumption, beyond the ground water recharge rate, reducing cropland rotation flexibility, increases in pesticide and fertilizer usage, and substitution whereby corn plantings displace other food crops thereby increasing other crop prices.

Some environmental groups also warn that increased use of nitrogen fertilizer requires using larger amounts of natural gas for nitrogen fertilizer production. Because crops do not absorb all the nitrogen, much of it leaches into streams and groundwater. That runoff has long been recognized as a major pollution problem, and it is growing. Vaclav Smil, a professor at the University of Manitoba, has calculated that without nitrogen fertilizers, there would be insufficient food for 40 percent of the world's population based on today's diets.

Are biofuels economic? Clearly in the near term they are not by virtue of the large federal subsidies required at least for the primary U.S. biofuel, corn ethanol. In the longer term it is unlikely that corn ethanol will ever be economic or competitive for several reasons. First, it is not a large net energy generator, and requires large energy inputs with little net energy gain. This is a major reason why corn ethanol production requires such vast planted acreage in comparison to solar thermal farms for the same energy output.

Burning liquid fuel in internal combustion engines only converts 17%-25% of the energy content into useful power as compared to electric vehicle engines, which convert approximately 80%- 90% including battery charging losses. We now see another of the reasons for the huge efficiency gains for solar electricity powered EV's over biofueled vehicles.

The Cellulosic Ethanol Mirage

It is useful to look at a comparison between what is considered an efficient future cellulosic ethanol production method using switch grass and using solar thermal (or PV) technology in terms of total energy produced per acre and total useful transportation energy produced. *The comparison is shocking.* For each acre per year, a solar thermal farm (17% assumed conversion efficiency) can produce 4500 Gigajoules of electrical energy versus 91 Gigajoules for the best assumed switch grass based futuristic ethanol.

The total energy conversion efficiency for turning the wheels of a car from the solar farm in California transmitting power over the HVDC grid to a New York EV car is 11.5% versus 0.125% for a high efficiency ethanol car using one of the most promising future cellulosic ethanol technologies. The economics appears to make no sense for switch grass or corn ethanol when compared to EV's and solar electricity. Even growing switch grass on marginal farm land for ethanol makes little sense in comparison to adding an additional 1/100 the land area of solar thermal or PV acreage in the southwestern arid uninhabited deserts or rooftops.

Understanding how much agricultural land would be required to be converted to cellulosic biofuel production is critical to understanding the practical limits of cellulosic ethanol production in the United States.

Waste product conversion to ethanol or biodiesel may make economic sense, since these resources would otherwise be discarded or under utilized. But the wholesale conversion of arable land to biofuels with such low efficiencies clearly is questionable public policy.

A comprehensive study under the auspicious of the bipartisan National Commission on Energy Policy titled, "Role of Biomass in America's Energy Future (RBAEF)", detailed what the land area requirements would be for transportation fuel conversion to cellulosic ethanol. This project was "unprecedented with respect to the breadth of technologies considered and the diversity of participants involved – representing the technical, environmental advocacy and policy communities," according to Lee Lynd of Dartmouth College in 2004, one of the project leaders and a key researcher on advanced biofuels.

RBAEF assumed that by 2050 there would be a need in the U.S. for 32 million barrels of oil for transportation in the United States. By assuming a doubling of the average light-duty vehicle fleet mileage to 50 mpg, the study assumed this consumption figure could be cut by 11 million barrels, leaving the need for 21 million barrels of oil for transportation.

RBAEF concluded that the U.S. has sufficient land to replace the equivalent of 8 million daily barrels with biofuels based on cellulosic ethanol in 2050.

RBAEF also focused on perennial switch grass crops and projected that its scenario would require 48-114 million acres of land, or 12-25 percent of current U.S. farm acreage to supply 8 million barrels per day, or 32% - 65% of all farmland to support the 21 million barrels to supply U.S. transportation needs in 2050 with cars achieving an average of 50 mpg. With the current mileage of the U.S. car fleet of 27.5 mpg, cellulosic ethanol would require 49%-99% of U.S. available farmland.

The equivalent land area in the unfarmed desert southwest would be approximately 1% of equivalent U.S. farmland area in 2050 not including efficiency gains from advances in solar farm conversion efficiency in the next 40 years and not including contributions from rooftop PV's, wind power or other renewable electricity generating capacity. This solar farm area would require little water, no pesticides, fertilizer, or transportation fuel expenditures to deliver the electricity to an electric vehicle transportation system.

How do the economics of U.S. biofuel subsidies compare with the solar electric/EV solution? In early 2008 the direct federal subsidy credit for corn ethanol was $0.51 per gallon. A May 3, 2008 Wall Street Journal article titled, "Corn Ethanol Loses Support", summarized the drawbacks to corn ethanol. A farm bill passed into law that same month by Congress raised the credit for cellulosic ethanol to $1.01 per gallon while reducing the corn ethanol credit to $0.45 from $0.51. The 2008 farm bill also had loan guarantees to build cellulosic ethanol plants and would pay farmers to experiment with biomass crops.

Let's compare the ethanol federal credits to the cost of electricity and solar electricity to power an EV. As an example, the Tesla is an EV sold commercially and can travel 220 miles on a 53 kilowatt-hour lithium ion battery. The average cost of electricity to fully charge the battery is assumed here to be 12.5 cents/kilowatt-hour, which is the average price of daytime electricity in 2008 for PG&E, California's largest electric utility. The result is that to "fill up" the Tesla battery and drive it 220 miles, the cost would be $6.62, or the cost equivalent of about 75 cents per gallon.

The federal 2008 <u>subsidy</u> for cellulosic ethanol is more then the entire mpg equivalent cost to charge up the battery of commercially sold EV's. It would save the U.S. Government money to simply give away the electricity as free fuel to EV owners compared to the cost of cellulosic ethanol or biodiesel subsidies.

But will the farm and ethanol lobby go along willingly with drastically reducing or ending the biofuels subsidies in favor of solar and EV credits? Not a chance.

If these equivalent subsidies were applied to electricity for electric bicycles and electric scooters in addition to EV's, large segments of the population could afford to commute to work and school with <u>zero fuel costs and zero air pollution</u>, paid by government subsidies, and taxpayers would still save money compared to existing and proposed new federal ethanol subsidies.

Corn ethanol also partially competes for corn used for food and feedstock. As the world's population strives to improve their diet and eat more protein, the demand for corn and other grain crops will only increase, thereby putting additional upward pressure on corn prices.

The unintended consequences of large scale corn ethanol production through subsidies and regulatory mandates are many-fold and include diversion of other cropland from food to fuel, higher food prices, depletion of grain stockpiles inevitably leading to famines and food shortages, overuse and depletion of ground water resources, increases in fossil fuel based nitrogen and potash fertilizer use and runoff into rivers and the ocean, and diversion of subsidy resources from economically viable and reliable solar farms and EV's that are 100% sustainable. In many air pollution districts, ethanol has also been found to be worse than gasoline as an air pollution source.

What is clear is that solar farm and rooftop PV electricity powered EV's will become far more competitive economically and environmentally than corn or cellulosic ethanol as a primary ground transportation solution, in addition to solar farms providing economic base load electricity.

The massive subsidies for current and future ethanol production are simply the result of misguided policy decisions and a fundamental flaw in the U.S. Constitution that gives disproportional Senate legislative power to low population farming states.

If the ethanol subsidies weren't foolish enough for their own lack of economic merit, looking at the CO_2 impacts of expanded biofuels production only shows how utterly foolish global biofuel policies have become. Indonesia is a good example of how countries can be induced to pursue dangerously ill-conceived energy policies. Indonesia continues to be one of the world's top three greenhouse gas emitters because of deforestation, peat land degradation and forest fires, according to a World Bank and British government climate change report "Indonesia and Climate Change: Current Status and Policies" released in 2007.

The report states, "Emissions resulting from deforestation and forest fires are five times those from non-forestry emissions. Emissions from energy and industrial sectors are relatively small, but are growing very rapidly". The report also states, "This may lead to harmful effects on agriculture, fishery and forestry, resulting in threats to food security and livelihoods."

According to the report, Indonesia's total annual carbon dioxide emissions stood at 3.014 billion tons. The United States was the world's top emitter with 6.005 billion tons followed by China at 5.017 billion tons in 2007.

Indonesia's yearly carbon dioxide emissions from energy, agriculture and waste are only around 451 million tons while forestry and land use changes are estimated to account for a staggering 2.563 billion tones according to the report. Indonesia's rainforests are being stripped rapidly largely because of palm oil plantations for bio-fuels, and illegal logging.

Many environmental organizations now expect Indonesia to lose most of its remaining rainforests and peat lands in the next few decades. If the reduction in greenhouse gas emissions was one of the goals from the promotion of biofuels, that policy has been a spectacular failure by all measures. It is estimated that it will take hundreds of years to recoup the amount of CO_2 released into the atmosphere that has been released from the burning of forest lands in Indonesia because of the use of palm oil based biofuels.

11

Peak Oil's Arrival

Some benefits from the approach of global peak oil production and increased use of EV's will be seen in unexpected ways. For many cities, like Los Angeles, California, the benefits will be seen most clearly through the arrival of cleaner and healthier air. As EV cars and trucks increasingly populate the freeways and roadways of Los Angeles, the air quality and city panoramic views will begin to return towards the air quality of the early 1900's. What was perceived as a crisis without end will be seen for what it really was, a new beginning era of cleaner, economic, more reliable energy.

This chapter will explain why peak oil is the reality of the 21st century, why we can be better off if intelligent policy decisions are made now, and the many benefits to be derived from the conversion to solar and renewable electricity, EV's and hydrogen fueled cars.

Peak oil is the point in time when the maximum rate of petroleum extraction is reached for a particular oil field, region, country or the entire planet, after which the rate of production enters terminal decline. As an oil field begins production, the rate of oil production generally increases until about ½ of the oil field's oil is produced.

After this peak has been reached most oil fields follow a predictable rate of decline with an ever decreasing rate of oil production. The production rates can be aggregated and applied to an entire country's aggregate oil production rate. Aggregating all countries production rates yields the total global production rate. Many geologists believe we have either reached or nearly reached this half-way point of maximum total global oil production for conventional oil resources. As more of the giant older oil fields are depleted at an accelerating rate, it becomes virtually impossible for new discoveries to offset the loss of production from the giant older fields.

If the world has indeed reached peak oil production for easily produced conventional oil resources, then the implications are profound. The billions of newly emerging consumers in Asia will be forced to compete for ever scarcer oil resources to fuel their new cars, plastic products and plane travel. Transportation vehicles and systems will have to become much more efficient and alternative fuels will have to be developed quickly in order to avoid much higher oil prices, recessions and future wars over increasingly scarce energy resources.

Ever since the geologist M. King Hubbert publicized his method for predicting when a particular oil field or country would experience peak oil, there has been controversy surrounding his theory. Whether the world has already seen or sees the peak in total global conventional oil production in the 2009-2014 timeframe has largely become a moot point. What is clear is that the world will not be able to continue to increase oil production to satisfy the growth in demand from the newly developing countries.

The increasing difficulty in finding large new inexpensive oil reserves and the accelerating depletion of existing oil fields are limiting the ability of the world's oil producers to expand production. Higher oil prices and movement to different substitutes will be the only way forward to balance supply and demand.

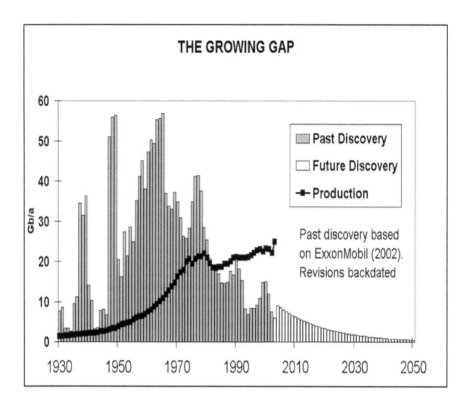

The world has been consuming far more conventional oil then it has been discovering for the past 20 years. This trend is unsustainable and we are beginning to see the impact of accelerating oilfield depletion rates. Enhanced recovery techniques (and recessions) have served to mask this mismatch in discoveries and production.

One myth that has been exposed is that Saudi Arabia will act as the world's oil swing producer and continue to expand oil production when needed. This world perception was fueled by Saudi Arabia's claim of more then 274 billion barrels of proven oil reserves and their claiming the capability to expand oil production to meet any global supply shortfalls.

Matt Simmons' book in 2005, <u>Twilight in the Desert</u>, detailed why the Saudi claims are a myth. Mr. Simmons studied many little noticed academic published reports on the problems in many of Saudi Arabia's largest oil fields, including Gwahar, which is the world's largest oil field. This field provides fully ½ of all Saudi oil production. Mr. Simmons came to the conclusion that the Saudi oil fields faced more severe challenges then the Saudi government was admitting in public. Saudi oil field production and depletion rates are considered state secrets by the Saudi government. The Gwahar field's production is over 60 years old and potentially facing irreversible production declines in the near term.

Advanced oil extraction technology has allowed increased production rates in many oil fields, but has also served to accelerate the depletion rates we are now facing. Many exporting oil countries like Norway, Britain, Mexico, and Indonesia have been seeing rapidly rising oil depletion rates. Indonesia has already become a net importer of oil due to oil field depletion. Many of the oil exporting countries are expected to become net oil importers within the next 10 years.

Mexico has been one of the United States largest oil suppliers and is now facing the collapse in production of the giant offshore Cantarell oil field in the Gulf of Mexico. Cantarell was the world's second largest oil field resource discovery, providing 63% of Mexico's total oil production. This oil field is expected to be nearly exhausted by 2015.

Mexico is now facing becoming a net oil importer as early as 2013 unless drastic changes are made to Mexico's deep water Gulf of Mexico oil exploration and development plans. Mexico's Constitution does not allow foreign ownership of Mexican oil fields and Mexico lacks the advanced technology required to find and exploit potential oil resources in the deep water areas of Mexico's Gulf of Mexico territorial waters.

Ironically, Mexico will likely embrace the same U.S. solar + EV mix because of its abundant sunshine and inexpensive land available for solar farms.

The trend is clear. The United States imports nearly 70% of its oil needs. The United States needs to use its existing fossil fuels more efficiently and rapidly develop its remaining oil resources in order to support the economic transition to solar, EV's and other renewables.

If the U.S. fails to effectively develop its remaining onshore and offshore oil and gas resources during this painful transition, the economic consequences will be even more severe.

The economic and national security risks of this foreign oil dependence places the country in a very precarious position as global demand for oil inevitably rises and depletion rates accelerate.

According to the U.S. Department of Energy, approximately 66% of U.S. oil consumption was used for transportation in 2008. Two-thirds of transportation fuel was for gasoline. The high price of oil in the last few years is accelerating the search for alternatives and causing a reduction in driving and suppression of demand for transportation fuels in many parts of the world. As new sources of transportation energy arise, the percentage of oil used for transportation energy will begin to decline from its current staggeringly high level of over 97%.

Each country's future energy needs and plans should be based in part on the needs of the country and international agreements regarding the environment. The best energy policy future for the U.S. will not necessarily be the same as for Japan, or Canada. It is becoming increasingly clear that each country which is a major importer of oil will continue to be faced with the need to reduce dependency on oil for economic and security reasons.

The solar and electric vehicle energy solution for the U.S. laid forth in this book takes into account the United States' unique resource environment. The solar farm electricity base load plus photovoltaics and EV solution mix will likely also be the best solution for many other countries that have the exceptional sunlight resources of the United States. The 94% perpetual availability in the Southwest, long range transmission potential, zero carbon emissions and pollution free characteristics makes it virtually ideal for the U.S. and other countries that have the similar solar resource.

The benefits of the movement to a combined solar + renewables grid and EV combination will be multi-fold and sustained. One of the additional risks we face with peak oil is the down slope part of the global depletion curve. As the world depletes the most productive oil fields and fewer new giant oil fields are brought online, we may be faced with an accelerating decline of available net exportable oil from the worlds' major oil exporters. Meanwhile, the oil exporting countries will see higher oil revenues over the long term, and will have less incentive to reduce demand growth for oil in their own countries, thereby reducing the net oil available to export.

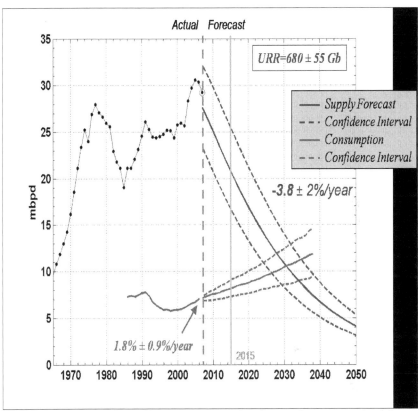

Estimated Top five oil exporters (production and consumption) by
Jeffrey J. Brown and Samuel Foucher

We no longer have the luxury of foolish energy policies driven by special interests. The faster we develop the best long term energy and transportation fuel resource alternatives, the less risk we will face of severe economic dislocations, job loss, potential famine and increased poverty.

Let's look at how the transition for the U.S and similarly situated countries away from oil import dependence can take place and what the benefits will be. Our dependence on foreign oil will not disappear overnight, or even in a decade. But once the right policies are in place, we can accelerate the transition with increased rapidity.

Let's look forward 30 years to see what the benefits will be from the solar grid + EV infrastructure and the move away from oil as the primary ground transportation fuel.

Energy security is the first benefit. There will be less need to protect shipping channels in the Middle East and less concern with the stability of oil exporting countries.

Oil producing countries will pay more for their own defense, since U.S. economic security will be less dependent on foreign oil. This will lead to a better trade balance through fewer oil imports, and the opportunity for a reduced military budget. We will also see increased economic competitiveness from the lower future cost of grid electricity for transportation.

Improved public health will be a direct result of reductions in air pollution, especially in urban areas. Drastic reductions in air pollution will result from electrification of transportation vehicles. Also, air quality will improve as a result of reductions in the burning of coal for electricity and reductions in the massive train loads of coal that are transported every day to power plants. Reductions in the use of diesel fuel required for trains that move the approximately 8.3 million box car loads of coal every year in the United States will also be substantial.

Because the world's largest consumer of oil will see decreasing oil demand, the price of oil may fall in the future, leading to the lower cost of goods and services that continue to rely on oil as a primary input. Countries that are not as oil independent will be less competitive. The U.S. will be able to return to economic expansion unconstrained by rising transportation fuel costs and energy shortages.

The electrification of small vehicles will serve to accelerate the electric transportation changeover. Electric bicycles, scooters, and mini-cars will make electric transportation more affordable to many people that would otherwise not easily afford gasoline fueled vehicles, or who currently rely on public transportation because of economic necessity.

China already sells more then 20 million electric bicycles per year to a population that cannot easily afford cars and would not be able to afford their high fuel costs. China currently does not achieve a maximum air quality benefit from its electric bike usage yet, since the increased electricity used is supplied largely from polluting coal power plants.

There will also be another significant global impact from the U.S. moving decisively to solar grid + EV energy. This will be a wholesale change in the U.S. position on greenhouse gas limits and regulation. Since it will benefit the U.S. competitively, the U.S. would likely become a stronger advocate of greenhouse emission caps globally.

Discontinuing or reducing subsidies and mandates for biofuels would free-up grain crops in the U.S. to be exported in much higher amounts to an increasingly hungry developing world. This will further improve the U.S. trade balance and increase our standard of living while reducing food price inflation. Since our transportation system will be using a much more efficient and cheaper fuel, the prices of EV transported goods should fall as well.

One of the unique features of the solar thermal + PV grid and EV roadmap is that people will have the knowledge and power in the near term to choose to largely limit their use of fossil fuels for direct energy and transportation vehicle needs. You can chose to buy PV panels for your home, or purchase renewable electricity through your electric utility and buy an electric vehicle. You have the power to rely much more on sunlight and other renewables for your power needs while helping to accelerate this transition.

It is also important to recognize that lessening the need for oil for ground transportation will serve to free up oil for use in those areas where it will be an essential fuel in the medium term, such as fuel for planes, trains and ships, and feedstock for plastics, drugs, and other essential industries where oil may continue to be the only practical choice for this century.

12

Stepping Off the Coal Train

Clean Coal Is Still an Oxymoron and More Expensive Than Solar Power

Coal plays a central role in the U.S. energy supply for electricity, providing 53% of the total electricity generated. The U.S. produces and consumes about one billion tons of coal per year. China consumes over two billion tons per year. The rising demand for coal for electricity generation and steel production in developing countries has resulted recently in historically high coal prices. The same limits on non-renewable oil resources will also begin to afflict coal. Burning thermal coal for electricity generation is also the largest contributor from human generated sources of CO_2 build-up in the atmosphere.

Coal is also a very large source of air and water pollution, acid rain, mercury and other toxins that carry great distances from coal power plants. Satellite imagery even shows that coal burning power plants in China are sending plumes of coal contaminated air across the Pacific Ocean and over the United States. Our oceans are also absorbing vast amounts of this pollution.

The primary reason that coal is the base load electricity power source for the United States and China is that both countries possess vast amounts of coal which is mined relatively cheaply. One of the main factors in the low cost of coal powered electricity is that the coal producers and coal power plants have been largely successful in avoiding paying for the environmental damage from burning coal. Ninety percent of all electric utility air emissions from CO_2, NOx, ozone, SO_2 (acid rain), and mercury are from burning coal.

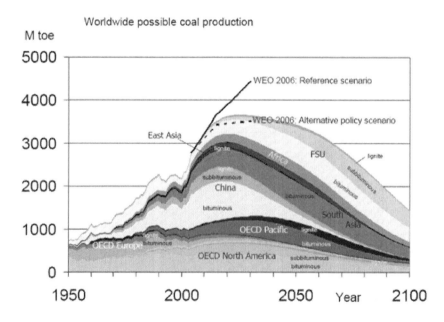

(Source: Energy Watch Group)

There are several factors that may cause the price of coal fired electricity to rise substantially over the next decade making solar generated electricity highly competitive with coal beyond the obvious environmental advantages of solar electricity.

As the reserves of coal in the eastern U.S. and Appalachia are declining, more coal is being sourced from the Powder River Basin in Wyoming and Montana. Most of this coal is then shipped by rail to Midwestern and eastern U.S. power plants. The costs of transportation and fuel have been rising making it more expensive to transport coal long distances. The price of Powder River Basin coal mined in the western United States and transported to east coast utilities was about $6 per ton in 2007. The transportation costs by train exceeded $20 per ton in 2008 to eastern U.S. power plants. As the price of future diesel fuel costs rise, the mining and delivery costs will also rise for coal.

The question of locating dozens more coal burning power plants in Wyoming and Montana and transmitting the power across state lines is not likely go very far for obvious reasons. The citizens of those states wouldn't stand for the increased pollution, sulfur, acid rain, mercury and water usage that would be required.

Various types of coal have different amounts of energy content. High quality anthracite coal production from Appalachia (30 megajoules of energy per kilogram (MJ/kg)) and Illinois--has been in decline since 1990. The U.S. "vast reserves" of coal frequently mentioned in the media are mainly lower-quality bituminous coal, delivering 18 to 29 MJ/kg, and sub-bituminous coal and lignite, delivering only 5 to 25 MJ/kg. As transportation costs rise and carbon caps or taxes are imposed on coal, the cost per kilowatt-hour will rise, making solar, wind and other renewable electricity much more competitive over the longer term.

For comparison purposes, the energy content of coal can be translated into "tons of oil equivalent." In terms of total volume mined, growth in the tonnage of U.S. coal production can continue for about another ten to fifteen years. But in terms of *energy*, which is the only metric that really matters, U.S. coal production peaked in 1998 at 598 million tons of oil equivalent, and fell to 576 million in 2005. Total coal energy content has risen recently, but is expected to peak within the next 10-20 years.

China's continued acceleration in coal consumption will inevitably result in a massive increase in imported coal demand from around the world, and higher costs for coal on the international market. Global coal energy equivalent production is expected to hit a peak around 2025.

Another of the major benefits from the conversion to significant base load solar power usage in the U.S. will be reducing the potential rise in coal prices due to rising Chinese and Indian coal demand.

Wyoming Ships More than 26,000 miles of Coal Trains per Year

It is likely that CO_2 taxes and carbon caps will eventually be imposed on burning coal which will result in higher prices for coal generated electricity. Many local and state governments have also refused to permit new coal power plants in their states due to environmental concerns. As demand for more electricity continues to grow, there will be more incentives for electricity sources that do not emit huge amounts of CO_2. Coal generated electricity emits about twice as much CO_2 per megawatt-hour as does natural gas.

There is much debate and discussion about moving to "clean" coal technology for power generation and even for coal-to-liquids for transportation fuels. One of the main goals of clean coal technology is to capture the CO_2 emissions from coal power plants. There are, however, no coal-fired power plants in commercial use which capture all carbon dioxide emissions, so the process is not proven as commercially viable and is still the subject of feasibility studies. It has been estimated that it will be 2020 to 2025 before any commercial-scale clean coal power stations will be commercially competitive, if ever.

It is, however, known that the cost of coal power plants with carbon sequestration would be very expensive if at all feasible. It will therefore be at least 10 years before any commercial scale clean coal power plants are running in the United States. Combining this timeframe and the very high cost of this technology and increasing cost of mining and transporting coal, it becomes clear that clean coal is an industry in search of a subsidized solution. Ultimately the economics of clean coal may not be competitive against the declining cost of solar thermal farms + thermal storage, wind farms and PV generated electricity in the United States.

One of the main demonstration clean coal projects supported by the DOE was the FutureGen project. FutureGen was a public-private partnership to build the worlds first near zero-emissions coal-fueled power plant. The 275-megawatt plant was intended to prove the feasibility of producing electricity and hydrogen from coal while capturing and permanently storing carbon dioxide underground. The project was canceled after the cost estimates rose above $1.8 billion. The estimated cost for the first clean coal power plant in the U.S. was already more then $6.5 per watt or over **two times** the peak power cost of currently contracted solar thermal farms from Pacific Gas & Electric in California.

The failure of the FutureGen "clean coal" project is a harbinger of why clean coal may be unlikely to compete effectively with solar farms for economically competitive electricity generation. The problem is that every ton of coal burned produces 3.7 tons of CO2. When taken to a national scale, in order to sequester the amount of thermal coal currently burned in the Unites States would require sequestering **3.3 billion tons** of

CO_2 per year. Solar farms and photovoltaics would require **zero tons** of CO_2 to be sequestered as a direct result of producing electricity.

Even if clean coal power plants were feasible and economic, consider the coal required to supply them. A typical coal train is a mile long and each hopper car holds 100 tons of coal which lasts only 20 minutes when burned in a coal power plant. Surface mines can load two or three mile long trains of coal a day. More than eighty trains leave Wyoming every day, or more then 26,000 miles of trains per year. The amount of fossil fuel burned to run these trains is enormous, and does nothing to reduce our dependence on foreign oil or fossil fuels.

Solar farms and PV's by contrast have zero fuel costs which also makes them very resistant to inflationary cost increases for delivered power over the next several decades.

Coal-to-liquids is another technology that has been much discussed as a possible partial solution to peak oil and higher foreign energy dependence. This technology was adopted by Nazi Germany during WWII because of an embargo on oil and Germany's significant coal reserves. This still widely used process is called the Fischer-Tropsch synthesis process and is currently used in South Africa, China and other countries. The process when applied to coal-to-liquid fuel production is expensive and results in twice the CO_2 generation as producing, refining and burning diesel fuel. Wholesale movement in the United States to a liquid fossil fuel technology that doubles the per gallon CO_2 emissions and is very expensive is unlikely to become popular in the U.S. and as you have seen is unnecessary.

Another reason there has not been a rush to build coal-to-liquid refineries is the extreme capital cost in plant construction. One plant producing 80,000 barrels of coal-to-liquid fuel a day would cost over $7 billion in capital outlays.   This cost is comparable to a 2.8 gigawatt solar farm which would provide more equivalent transportation energy miles then the 80,000 barrel per day coal-to-liquid plant. The solar farm would also pay nothing for its energy source, whereas the coal-to-liquid plant would pay between $140 million and $420 million per year in coal costs.

Even without the cost of the carbon emissions, water or pollution costs, coal-to-liquids is not competitive on a transportation energy or mileage equivalent basis with a solar farm and electric vehicle infrastructure.

As more and more states deny new coal power plant permits, it may be even less likely that states will permit significant numbers of coal-to-liquid facilities. There are proposals to capture carbon emissions from the coal-to-liquids facilities for use in oil field recovery enhancement. Whether these facilities will ultimately be economically successful is still doubtful.

The dynamics between permitting new coal power plants and the growth of solar thermal farms integrated with natural gas backup cogeneration and rooftop photovoltaics will rest largely on economics, changes in national electric grid policy, and the level of CO_2 emissions caps, taxes and regulations.

As solar energy use grows and declines in terms of cost per kilowatt-hour, there will likely be greater resistance to permitting new coal power plants. If solar farms and PV's decline sufficiently in cost, there may even be pressure to close older coal power plants as solar power with overnight storage takes a larger share of the base load power.

Wind power will also play a role in this transition. The DOE has estimated that as much as 20% of U.S. generating capacity may come from wind power by 2030. T. Boone Pickens, the billionaire oil maverick, has an ambitious program to invest billions of dollars in Texas wind farm electricity production and displace a portion of domestic natural gas used to generate electricity.

The extra natural gas availability would allow it to be used as domestic compressed natural gas transportation fuel and displace up to a third of U.S. imported foreign oil by 2020. T. Boone Pickens' plan highlights the inevitable economic shift to solar and wind generated electricity.

The greater use of natural gas as transportation fuel in the U.S. can indeed lower foreign oil imports. T. Boone's plan is well grounded in economics and national security concerns, and is environmentally sound. The publicity generated by T. Boone's wind power plan was only one of the opening salvos of the much larger shift to solar and wind generated electricity in our near future.

13

Reaching the Solar Tipping Point

The United States and the world's perpetual solar energy powered future will become a reality. As you have seen, increasing investments in solar energy usage are a path to a more secure clean energy infrastructure. Solar energy use needs to be supported both through smarter government policies and individual choices. Hopefully this book has made some of the policies that will lead to a more intelligent, safer and secure solar powered future more clear.

This chapter addresses how to take advantage of the growing availability of solar resources and electric vehicles and summarizes the policy choices that need to be made at the local, state and federal level.

Solar thermal farms with natural gas cogeneration combined with photovoltaics satisfy the 5 criteria (attributes) previously set forth that are required for a future base load electricity and transportation fuel source. Large scale national utilization of solar energy can be cost-effective, safe, scalable, reliable and does not have adverse consequences.

The future increase in electricity demand that will be required to support economic growth and fuel the rising numbers of electric vehicles can be entirely supplied, at least in the case of the United States, with increased solar and wind powered electricity. The near perfect combination of solar generated electricity and EV's is an opportunity for this generation to lay down the infrastructure for a permanent long term sustainable, pollution free power source that will free the U.S. and many other countries from the depleting and insecure foreign supplies of foreign energy. Freedom from OPEC and other oil exporting countries' influence and their loss of control over transportation energy prices will be seen as a blessing by future generations.

Domestic oil and gas resources will still need to be pursued and produced with vigor over the next 20-30 years in order to help pay for the renewable energy transition. The massive dependence on foreign imported oil is one of the greatest remaining dangers to the U.S. ability to successfully transition to renewable energy and reduced foreign oil dependence while avoiding major economic dislocations.

Increasingly fuel efficient transportation vehicles will help limit foreign oil dependence. But the increasing depletion of domestic oil resources will still require sustained support for domestic onshore and offshore oil and gas production throughout this extended renewable energy transition period.

Depending on where you live, you may be able to choose solar energy for three of your main types of energy usage. If you begin to use solar energy for your electricity, water heating and driving, you will have started well down the road to becoming less susceptible to future rising costs of fossil fuels. You have many solar related choices for each of these three energy uses.

With respect to electricity, you can choose to install photovoltaics where you live or purchase supplemental renewable utility grid electricity through a certified renewable electricity provider that is from multiple renewable sources or from solar generation.

If you choose to install PV panels you will also be buying your home or business a transportation fuel source for when you purchase electric vehicles. Companies like Solar City can even finance the installation of PV's for your home or business and arrange a lease-back plan for the electricity. Some of these plans require no upfront money and pay for themselves over time. As the price of solar panels drops, you will be able to lease your rooftop space to a solar integrator and immediately save money on the cost of your electricity.

You will be buying significantly increased independence from foreign oil sources, reduced fossil fuel dependence and limit your future energy cost increases. You will also be setting an example for friends, businesses, neighbors and your community.

As more and more people and organizations sign up for solar electricity and as states mandate more renewable energy usage by utilities, more solar farms will be built. This will result in less demand for nuclear, natural gas and coal by utilities and independent power producers.

If you live in a sunny climate or a climate where solar water heaters can be effective, then you can check with your local solar water heater installation companies and get a quote. As fossil fuels inevitably become more expensive, you will see your payback time decrease.

Choices for electric vehicle purchases will grow rapidly over the next few years. China has already moved to ever larger sales of electric bicycles for its people. China's largest cell phone battery maker, BYD, is already producing and selling plug-in hybrid electric cars that can be driven over 60 miles on battery power alone.

There are many plug-in hybrid cars that will be on the market in the next few years. For commuting to work and back, they offer the opportunity to charge their batteries with solar and renewable electricity and drive pollution-free. Tesla, Think, Miles, Phoenix Motors, ZAP and other pure electric cars are already being sold and can be "fueled" with electricity. Corporate fleets and businesses are buying electric utility vans in Europe. Corporate fleets of electric vehicles will be accelerating in the next few years. Heavy lift 100% electric trucks are already under development and will be garnering increasing worldwide sales.

Companies are beginning to offer conversion of standard gasoline cars to 100% electric. Consumers have the option of replacing traditional engines in their cars with 100% electric engines.

As an example, a company is offering a conversion package for the Saturn Sky which can travel 150 miles on a single EV battery charge, a top Speed of 90 mph and 0-60 in 5.7 seconds.

The AMP Saturn Sky

The combustion drive train is replaced with electric motors, lithium batteries, and electric control software. Cost of this conversion to 100% electric runs about $25,000, plus the cost of the car, making the entire total about $50,000. The price of these conversions will only drop over time.

Some electric cars can be charged in under 10 minutes for EV's with batteries using newer technology and charged at special charging stations with more stations being built every week. In most states, the cost to charge these cars is around 60-75 cents a gallon equivalent. You can also choose to buy an electric scooter, motorcycle or bicycle for local commuting, school, or just for fun.

Chinese made electric cars are also being sold in the U.S. The Flybo is a small sized two passenger electric car with a top speed of about 40 mph. Jinan FlyBo Motor Co. of China produces the car. It can run 150 miles before requiring a four-hour charge using an ordinary household 25 amp outlet. A 48-volt electric motor powers the rear-wheel drive FlyBo, which has a 71-inch wheelbase and seats two and costs about $13,000.

With the changeover to electric vehicles powered by renewable electricity you can also slash your fuel costs. You can begin to stop thinking in terms of consuming a depleting resource, but rather in terms of a 100% renewable, clean and cheaper transportation fuel source.

You will also be able in the future to choose to enable your electric car battery to serve as a backup electric power source for your house and be more secure against power outages.

Limiting Future Fuel Price Increases

The driving distance from Los Angeles to San Francisco is 370 miles. With an electric car, the trip would cost about $20 in fuel roundtrip, assuming 12 cents/ kilowatt-hour for electricity. Even using pure renewable electricity at 16 cents per kw-hour, the cost in fuel would only be $28 roundtrip. The direct subsidy cost alone to tax payers for an ethanol fueled car on the same trip would be over $29 not including the cost of fuel to the consumer at the gas pump.

Walking, bicycling, car pooling and taking public transportation are also good choices for getting from one place to another. But technology does offer more choices. Informed choices are important. Through purchasing renewable electricity or installing PV's at home, combined with electric transportation and using solar water heating, you have eliminated the vast majority of your fossil fuel energy use.

Whether or not most global warming is man-made and how serious are the near term issues of man-made CO_2 build-up will eventually be better understood.

Corn ethanol, soy biodiesel and other food crop fuels are poor policy choices for our food supply and it makes neither economic sense nor environmental sense to promote or buy food crop ethanol in the United States. These crops also require fertilizers produced typically using natural gas. Just remember that the more you fill up your gas tank with corn ethanol, the more food you've burned.

Write a letter or email your congressional representative and ask them to reduce mandates for ethanol use and subsidies for ethanol, nuclear and coal usage. Bring these issues up at your next congressional town hall meeting.

Ask your local environmental groups to drop support for food crop ethanol mandates and subsidies, and turn their efforts to solar and wind generated electricity and electric vehicles instead.

Tell your local, state and congressional representatives that you favor more solar energy over new applications for coal power plants.

Loan and insurance guarantees for the nuclear industry for power plant construction, operation and accidents are not really part of the free market system. They are simply subsidies to another special interest group. If there is a hearing scheduled in your community regarding a nuclear permit application or Environmental Impact Review, attend and speak your mind about new nuclear licenses. The next nuclear accident or disaster may not be as benign as Three Mile Island.

U.S. oil companies received more then $18 billion in tax breaks and subsidies from the Federal government in the last several years. Support more solar energy incentives, rebates and tax breaks. The result will be an acceleration of new solar power generation capacity, innovation and investment and more rapid production of EV's.

Support increased interim onshore and offshore domestic exploration and production of oil and gas. It will take many years for the U.S. economy to lose its foreign oil dependency. Maximizing domestic oil and gas sources in the meantime is an important way for the country to minimize its enormous trade deficit and security risks due to foreign imported oil.

The solar industry simply hasn't achieved the same level of political support for solar energy's fundamental merits compared to other lobbyist support groups from farm state senators and representatives that are willing to degrade America's economy and energy security for their special interest ethanol constituents.

The growth in demand for solar electricity and EV's can progress rapidly over the next two decades. Western states will likely lead the growth of large solar farms initially with daytime solar thermal and PV generation which is already underway in California, Arizona and Nevada. Solar thermal farms and photovoltaics will reduce or eliminate the need for more fossil fuel power plants first in the Southwest.

PV installations will grow in all parts of the United States. As PV prices decline, we will see ever more rapid adoption by homeowners, utilities and businesses.

As solar generation capacity grows, there will be more and more daytime electricity available. Solar thermal farms will then begin adding thermal overnight storage and begin to displace even regional nighttime electricity generated from coal and natural gas. As the rest of the country sees the success of solar farms, photovoltaics and wind farms' ability to generate pollution free and economic electricity, the opportunity to send power over a more efficient HVDC grid to the entire country can become a reality. Tax credits and incentives for solar electricity may eventually accelerate the adoption and growth of large solar farms becoming national suppliers of electricity for day and nighttime usage. One very large advantage of solar thermal farms in the U.S. is their use of domestically produced components and materials.

The accelerating usage of plug-in hybrids and pure electric cars will increase the demand for electricity. With the right regulatory framework, the southwestern solar farms and countrywide rooftop photovoltaics will begin to serve the increase in electricity demand from the EV market.

A significant challenge is to understand how your choices can move the country and the world down this safer, cleaner and smarter solar renewable path. We all can address this challenge at the local, state and national levels.

At the local level, it is important to be informed regarding new permit applications for local power plants. It is a common refrain that new fossil fuel or nuclear power plant permits must be approved because solar and other renewables will not be able to supply the power needs for future growing electricity demand. As we have shown, these claims are not accurate.

It is important to understand the business interests that promote a particular fuel. In the case of coal powered electricity, business interests include the coal mines, coal mine equipment suppliers, railroads, power plant builders and operators, air pollution equipment industry, and turbine manufactures. The coal industry also has encouraged coal power plant approvals on the premise that the more modern designs can retrofit CO_2 sequestration capabilities. Carbon sequestration for U.S. power plants has not been demonstrated in large commercially competitive settings and is much less mature then the current generation of commercial solar thermal farms.

The nuclear industry has supporters that promote nuclear power plants with very large hidden subsidies. The nuclear power industry interests include uranium miners, refiners, enrichment, nuclear power plant designers and builders, and companies under contract for Yucca Mountain waste storage and management.

Local level incentives should be supported for solar energy for individual homes, businesses, and local government projects. Longer term, there is no doubt that solar energy will be economically competitive on a level non-subsidized playing field, and with zero pollution when generating electricity.

Electric vehicle charging stations should be encouraged and receive favorable zoning. Electric car parking preferences and right of ways can be encouraged. If the United States can provide billions in military expenditures for security for other nation's oil routes and facility protection, we can afford incentives for electric vehicle charging stations and renewable electricity credits that move us to domestic, pollution-free energy independence.

At local town hall meetings, county supervisor meetings, city counsel meetings, environmental working group meetings, conferences and with elected officials, express your concerns regarding corn, soybean or other food based ethanol promotional programs. Question residents, businesses and government agencies regarding their support for food based ethanol for fuel at the expense of solar incentives.

It is also surprising that many city and county governments not only do not encourage solar photovoltaic installations, but actively discourage or forbid solar PV installations where they can be seen from the street. These antiquated policies need to change.

It is also important to support zoning and permitting applications for solar farms and high voltage right-of-ways that support renewable energy and solar farms.

Solar farms + photovoltaics are capable of producing and supplying much of the electricity required for nearly all regions of the country, day and night. Members of the House of Representatives should hear from their constituents the reality of what havoc the food crop ethanol farm subsidy programs have caused on food and fertilizer prices, and House and Senate members need to stand up for the needs of the majority of the population that pay the taxes for these inefficient policies.

For any remaining doubters about the United States ability to convert to a large scale 24/7 solar power and electric transportation economy, ask the following question. Is it really credible that a country that developed the technology to return men from the moon safely, builds the world's most advanced supercomputers, manufactures the worlds most sophisticated jet engines and planes- that this country can't design and build solar farms that do nothing more then focus sunlight on heat pipes that store heat and generate steam that runs turbines which generate electricity and transmit the power over modern long distance electric power lines to battery driven cars?

America and many other countries have an abundance of sun drenched uninhabited and available lands and roofs to build the solar farms and photovoltaics that can power our long term energy future. Is there really anything stopping us from reaching the solar tipping point?

Notes

Introduction

4 **It is well documented:** Science News March 11, 2007

4 **March 2006 edition of Scientific American**, "The Dangers of Ocean Acidification," by Scott Doney, one of the authors of the study published in Nature.

5 **The sheer scale:** Fuel from the Sky: Solar Power's Potential for Western Energy Supply. http://www.nrel.gov/csp/pdfs/32160.pdf

Chapter 1 The Current Energy Crisis

11 **Oil consumption in the U.S. grew:** 1979. United States Control of Petroleum Imports, page 12. By Torleif Meloe

Chapter 3

32 **Solar trough or linear array**: Oct. 2003 Report. NREL/SR-550-34440 Assessment of Parabolic Trough and Power Tower Solar Technology Cost and Performance Forecasts.

32 **In 1910, Frank Shuman:** A Golden Thread, 1980. Ken Butti and John Perlin.

38 **The Spanish renewable energy:** April 1, 2008 Associated Press New Release.

41 **In the eastern California deserts**: Edwards AFB Ecosystem program. http://www.mojavedata.gov/edwards/contacts.html

49 **The 2008 estimated U.S. corn planted:** USDA March 31, 2008 Crop Survey Release, "**USDA Expects Corn Acres to Drop in 2008 as Soybeans, Wheat Gain Ground**"

49 **Comparatively, solar farms**: 2008 Ausra Study, SOLAR THERMAL ELECTRICITY AS THE PRIMARY REPLACEMENT FOR COAL AND OIL IN U.S. GENERATION AND TRANSPORTATION.

49 **If the U.S. were to move to solar farms:** Assumes 1.8 trillion passenger vehicle miles driven per year (US DOT estimate), 3.1 miles/kWh for EV passenger car, 1,230 MWH per day per solar farm square mile (7.2kWh/day/m^2 solar radiation (eastern Mohave Desert, CA)) equals ~1,150 square miles.

50 **Widespread geographic**: Global Status Report 2006 Update. http://www.ren21.net/globalstatusreport/download/RE_GSR_2006_Update.pdf

50 **Widespread geographic:**
http://www.inhabitat.com/2008/01/14/energy-breakthrough-storing-solar-power-with-salt/
53 **Unfortunately, the U.S.:** Narain G. Hingorani in *IEEE Spectrum* magazine, 1996.
53 **HVDC:** http://en.wikipedia.org/wiki/Hvdc
53 ABB Group 2008
http://www.abb.com/cawp/db0003db002698/02de19e1cb36dbd4c12572f4004552fb.aspx
53 **The electrification of our transportation system:** April 21, 2008 Canwest News Service, "Transport energy now 'No. 1 issue' facing U.S." 53 March 2008. R. Keith Evans. "An Abundance of Lithium", R. Keith Evans & Associates.

Chapter 4

67 **Rooftop leasing and PPA's:**   U.S. Energy Information Agency, 2007 Household Usage Report.

Chapter 5

90 2000 California Motor Vehicle Code 406(a)

90 California Vehicle Code Division 11 - Rules of the Road. Chapter 1. Obediance to and Effect of Traffic Laws. Article 5. Operation of Motorized Scooters

Chapter 6

102 **Solar thermal storage technology:** A Golden Thread, 1980. Ken Butti and John Perlin., p. 100
104 **In previous U.S. Government demonstration programs:** Sandia National Laboratory Report.
http://www.sandia.gov/Renewable_Energy/solarthermal/NSTTF/salt.htm

Chapter 7

113 **The first modern commercial power:**
http://en.wikipedia.org/wiki/HVDC

Chapter 8

121 **Solar water heating on Spanish buildings:** Spain Solar Hot Water Ordinances: Source: "Codigo Técnico de Edificación," Real Decreto 314/2006 + Documento Básico HE.

Chapter 9

131 The **cost alone for the only U.S. high level:** 1995 National Academy of Sciences Study
131 2000 DOE Yucca Mountain Cost Estimate

Chapter 10

142 **Understanding how much agricultural land**: "Ethanol Timeline" produced by the U.S. Department of Energy's Energy Information Agency.
http:/www.eia.doe.gov/kids/history/timelines/ethanol.html

141 **Some environmental groups also warn:** 2005 Tad W. Patzek and David Pimentel, Natural Resources Research (Vol. 14:1, 65-76)
141 2007 Corn-based ethanol- A Case Study in the Law of Unintended Consequences, Competitive Enterprise Institute.
141 **NY Times** Shortages Threaten Farmers' Key Tool: Fertilizer Wednesday April 30, 2008.
144 **In many air pollution districts:** "Effects of Ethanol (E85) Versus Gasoline Vehicles on Cancer and Mortality in the United States," by Mark Z. Jacobson, Stanford Univ., April 18, 2007 online edition of *Environmental Science & Technology*
144 Lee R. Lynd, Testimony Before the Senate Committee on Agriculture, Nutrition and Forestry Hearing on Emerging Opportunities for Utilizing Agricultural Biomass to Enhance Future Energy Production and Security, May 6, 2004.
http://www.climatesolutions.org/publications/CS_Growing_Sustainable_ Biofuels__The_Place_of_Biofuels_2007-07-16_26.pdf

Chapter 11

153 Published on 7 Jan 2008 by GraphOilogy / Energy Bulletin. Archived on 8 Jan 2008.

Chapter 12

161 **There is much debate and discussion:** David Brockway, Chief of the Energy Technology Division, CSIRO, quoted by Crikey.com.au 20 Feb 2007

163 **The DOE has estimated:** May 12, 2008. DOE Report. "20 Percent Wind Energy by 2030"

160 **China's continued acceleration:** China: Coal to Oil in 2008. Published by Andy Rowell September 17th, 2007 in China, coal to liquids. Oil Change International.

163 28 Mar 2007 by Energy Watch Group Coal: Resources and Future Production.

INDEX

voltage   113-4

W
water
  electrolysis   103, 107
  heating   121, 123, 166
western United States   159
Westinghouse, George   110-1
wind   15, 50, 84, 93, 103, 106, 117,
        159, 163, 166, 170
  farms   107, 109, 115-6, 161, 171
  power   4, 5, 16, 18, 27, 41-2, 118,
        143, 163
wood   10

Z
ZAP   87-8, 167

Made in the USA
Charleston, SC
14 February 2010